AQUACULTURE SECTOR PLANNING AND MANAGEMENT

COLIN E. NASH

Former Programme Leader
Aquaculture Development and Coordination Programme
Food and Agriculture Organization of the
United Nations, Rome, Italy

Fishing News Books

© 1995 by Fishing News Books
A division of
Blackwell Science Ltd
Editorial Offices:
Osney Mead, Oxford OX2 0EL
25 John Street, London WC1N 2BL
23 Ainslie Place, Edinburgh EH3 6AJ
238 Main Street, Cambridge Massachusetts 02142, USA
54 University Street, Carlton Victoria 3053, Australia

Other Editorial Offices:
Amette Blackwell SA
1, rue de Lille, 75007 Paris
France

Blackwell Wissenschafts-Verlag GmbH
Kurfürstendamm 57
10707 Berlin, Germany

Feldgasse 13, A-1238 Wien
Austria

First published 1995

Set by Best-set Typesetter Ltd., Hong Kong
Printed and bound in Great Britain by
The University Press, Cambridge.

DISTRIBUTORS
Marston Book Services Ltd
PO Box 87
Oxford OX2 0DT
(*Orders*: Tel: 01865 206206
Fax: 01865 721205
Telex: 837515)

USA
Blackwell Science, Inc.
238 Main Street
Cambridge, MA 02142
(*Orders*: Tel: 800 215-1000
617 876-7000
Fax: 617 492-5263)

Canada
Oxford University Press
70 Wynford Drive
Don Mills
Ontario M3C 1J9
(*Orders*: Tel: 416 441-2941)

Australia
Blackwell Science Pty Ltd
54 University Street
Carlton, Victoria 3053
(*Orders*: Tel: 03 347-5532)

A catalogue record for this title is available from the British Library

ISBN 0 85238 227 8

Library of Congress
Cataloging-in-Publication Data is available.

Dedication

I dedicate this book to my wife, Patricia, and my family, who have patiently endured my association with aquaculture for over three decades, and graciously accepted without murmur all the trials and tribulations (and all the pleasures) of moving and living in so many different places around the world as I worked with this emerging modern industry through its most exciting years.

Contents

Preface

My primary purpose for producing this book on *Aquaculture Sector Planning and Management* is to fill the great void for fundamental information needed by planners and administrators in government charged with the responsibilities of organizing and managing the relatively new and important food-producing sub-sector, now widely called aquaculture. Secondly, I would like it to be a useful guideline on planning and management processes for anyone in the public or private sectors who has some responsibility for planning and implementing aquaculture development programmes or projects. Finally, I hope it is a different type of textbook for undergraduate students, including those in the closely related fields of agriculture and environmental sciences, on the many issues underlying the development of aquaculture, or that of any other similar field.

As the book is basically a guideline I have not attempted to prescribe model methodologies for planning, simply because there are too many major differences between countries in their legal, economic, and social frameworks. Rather, I have tried to identify and describe the principles basic to the creation of a successful aquaculture sector, and then to explain the necessary considerations and actions for successful organization and management, and developmental planning. Interspersed and highlighted throughout the text are examples which illustrate how individual countries are approaching development issues and meeting sector needs. Although some of the illustrations may in fact now be dated, such as references to organization within countries with centrally-planned economies or quantified duties and tariffs on fisheries products, they nonetheless still serve as real and relevant examples to the points made in the text.

I have divided the book into three parts. The first part (Chapters 1–4) concerns Aquaculture Sector Planning. The first chapter provides an overview of the sector. It identifies a framework of all its diverse forms, and briefly describes the many systems and practices available for producing aquatic animals and plants. This is followed (Chapter 2) by summarizing the many environmental, social, and market factors which influence the sector's growth. Chapter 3 deals with the sector planning process, describing the many components of planning, and the problems, before ending with a description of regional planning. The first part concludes (Chapter 4) with the practical aspects of preparing a national sector plan, and the formulation of strategies for development.

The second part (Chapters 5–9) deals with aquaculture sector management. It begins (Chapter 5) by identifying the roles of the public and private sectors in management, and their separate responsibilities, followed by the typical organizational frameworks in which they operate. It ends with the practical issues of organization and management, specifically identifying the management tools available to any government. Chapters 6, 7, and 8 describe in detail the three principal management tools, namely the policy instruments of legislation, fiscal measures and support services, and development projects, respectively. The second part ends (Chapter 9), with a description of the fourth important management tool, namely monitoring and evaluation.

The final part of the book (Chapter 10 only) concerns project planning and management. It is not totally relevant to sector planning and management as projects are a lower level of activity, but it has been included because development and investment projects are policy instruments used for sector management (Chapter 8). The chapter should therefore be considered an Annex only, providing briefly another planning framework useful to government administrators and entrepreneurs considering new projects.

The chapters are intended to be both informative and instructive. Through them I have attempted to provide a complete review of all aspects of the aquaculture sector, using selected illustrations from countries worldwide to emphasize specific points. Some chapters also include conclusions on, for example, the practical issues of planning, organization and management, and selecting and implementing policy instruments for fiscal measures and support services. The last chapter is primarily instructive in intent.

The planning behind the preparation of this work, and some of the preliminary writing, was carried out when I was Programme Leader of the Aquaculture Development and Coordination Programme (ADCP), an interregional project of the United Nations Development Programme (UNDP) headquartered in Rome at the Fisheries Department of the Food and Agriculture Organization (FAO). My objective at that time (1985–90) was for ADCP to focus only on information for sector planning and management of aquaculture, rather than the dissemination of technical information which had been the previous emphasis of ADCP and that of the Regular Programme of FAO. Some of the results are evident in the ADCP documents listed at the end of this book, many of which emanated from preparatory studies for a comprehensive work on Aquaculture Sector Planning and Management to be published under the auspices of ADCP and FAO. Unfortunately, the ADCP project was terminated by UNDP in 1990, after 15 years of support, and there the effort ended for lack of financial resources. I am therefore indebted to FAO for permission to allow me to complete the task on my own, using the materials that had been assembled, and quoting generously from the original ADCP reports and documents commissioned for the purpose. Also I acknowledge the generous contribution of FAO's Photographic Library of several illustrations for the text.

The idea and structure for such a book was developed in close cooperation with David Insull, of the Fishery Development Planning Service, FAO. Consequently, I owe much to David for his continuous personal and professional interest, collaboration, and review of the various chapters. Other professional inputs, with the preparation of baseline materials and innumerable hours of general discussion, were made by several other colleagues and associates to whom I am also indebted and readily acknowledge. These are Mr Edward H. Nichols, Ms Annick R. Van Houtte, Ms Nicola Bonucci, Mr William R. Edeson, Mr Neil J. MacPherson, Mr P. A. D. (Paddy) Secretan, Dr Michel Girin, and the late Dr J. Anthony Mollett. In particular I also acknowledge the critical editorial review provided by Dr H. Francis Henderson, ex-Director of the Fisheries Resources and Environment Division (FAO), and his continuous encouragement to get the work completed and published. All this significant help notwithstanding, I still must take final responsibility for the statements and opinions expressed in the text, and do not imply that they reflect the views of any one of these individuals, or their organizations, including FAO.

I also recognize the generosity of Elsevier Science Publishers B.V. (Amsterdam), which permitted me to use some basic material which I had authored in previous works already published.

Finally, I would like to point out that two people who helped significantly in the evolution of this work, namely Tony Mollett and Ted Nichols, drew on their considerable experience in planning, and organization and management, in agriculture. Neither had any previous knowledge of the aquaculture field, but rapidly became experts. Tony Mollett in particular commented many times on the close parallel of the sector planning processes for both agriculture and aquaculture, because of the use of the same national and natural resources, and the same social and economic constraints to development. I regret he did not live to see the finished work published but I think he would genuinely support my hope that the book has a general appeal and value to many teachers and students of planning in most of the other sub-sectors of agriculture, to which aquaculture appropriately belongs.

Colin E. Nash
Rolling Bay
Bainbridge Island
Washington, USA
5 November 1994

PART I
AQUACULTURE SECTOR PLANNING

Chapter 1
The Aquaculture Sector

Synopsis

The chapter begins with a description of the current trends in aquaculture production world-wide, and notes the importance of statistics as a basis for planning and investment in this growing sector. There is a brief comment on the lack of an acceptable definition of what aquaculture is, before the main section describing a framework to categorize the diversity of the sector, and the various management approaches to production due to the economic, social, and cultural factors behind the interventions by man in the life-cycles of aquatic animals and plants. After a section on the systems and practices of aquaculture technology and species preferred for culture, the final section deals with other external components which are important for forecasting growth. This broad and general introduction is intended to indicate the complexity of the decision-making processes for any particular country administrators or group of people planning or investing in the sector.

1.1 Aquaculture production

Aquaculture is a relatively new industry for the production of marketable commodities, mostly food for human consumption. In spite of its ancient origins in the traditional crafts of aquatic plant and animal husbandry in Asia, modern aquaculture, with organized management under-pinned by research and development, has only emerged as a viable economic and industrialized sector within the last fifty years.

The measure of annual production and growth of aquaculture in any country is important to planners. It not only guides them in quantifying and qualifying the scope of any planning effort, but it also demonstrates the continuous progress or success of the previous plan, or points out any weaknesses. Planners also need to know what other countries are producing, as this may indicate new market opportunities, better technological applications, or better planning.

The responsibility for the collection of aquaculture statistics so far has rested with the United Nations Food and Agriculture Organization (FAO).

The majority of all FAO's annual statistics are supplied by FAO-member governments. Although most governments, through the appropriate department, conscientiously assemble correct data before submitting them, some governments are not as well informed as they might be about their sectors, particularly if the sector is new – such as aquaculture. Sometimes either no data are reported at all, previous data are repeated, or the data are merely estimates based on what was provided the year before.

Figure 1.1 summarizes the growth in global aquaculture production in the last ten years, and its increasing importance in the world harvest of fisheries. It includes only the major groups of aquatic animals (finfish, crustaceans, molluscs, and some miscellaneous animals), but does not include the considerable production of aquatic plants (mostly marine algae) which are farmed annually.

Global data were assembled for the first time in 1976 for the FAO Technical Conference on Aquaculture, held in Kyoto, Japan. Some 67 countries were identified as producers for that occasion, with a combined total production of just over 6 million metric tons (t). This was made up of some 4 million t of finfish, about 1 million t each of molluscs and marine algae, and less than 16 000 t of crustaceans. The countries of Asia combined to produce over two-thirds of the total, led by China which alone produced some 2.5 million t. Excluding the production of aquatic plants, which are not included in FAO's annual fisheries statistics, aquaculture production contributed less than 7% to the global fisheries harvest.

Data are now collected annually on a formal basis by FAO, from information supplied to them under conditions described above. Global pro-

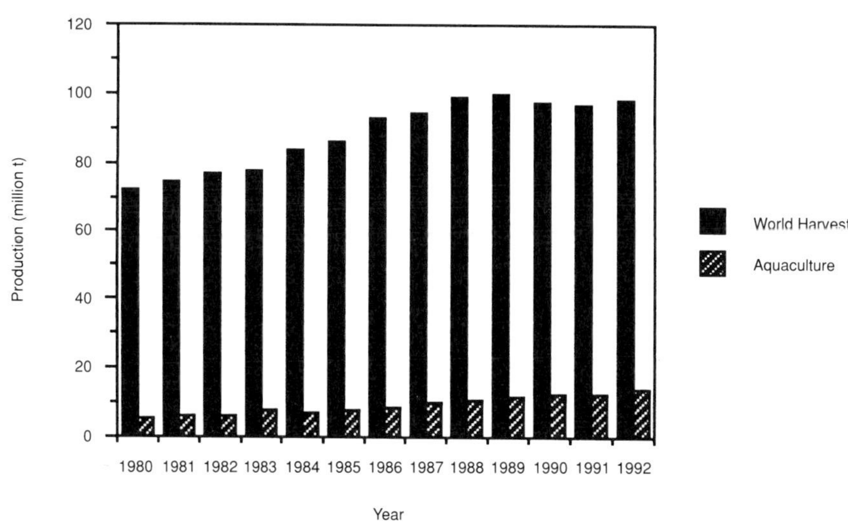

Figure 1.1 Global aquaculture production, 1980–1992, (in million t) as a proportion of the total world harvest of fisheries (Source: FAO, 1994).

duction has increased continuously since then and now surpasses 13 million t. Again excluding aquatic plants, production is over 15% of the global harvest. Finfish continue to make up two-thirds of the total, but the high value of crustaceans on world markets has been the reason for large investments in shrimp and prawn farming, and production is now over 800 000 t. Molluscs and marine algae continue to grow steadily, and annual production is about 3 and 4 million t, respectively.

The value of aquaculture production was estimated for the first time in 1984, when it was about US$12.5 thousand million. It is now well over US$30 thousand million. This is about one-third the value of the total world fisheries catch, due to the higher values of aquaculture products for human consumption compared with low values of large volumes of capture fisheries which are reduced for fish meal and oils.

For the most part the substantial increases in farmed commodities have contributed significantly to meet increasing demands for fisheries products, estimated at 120 million t in the year 2000. Moreover, almost all animal aquaculture commodities are used directly for human consumption. Conversely, about 29% of the world catch is not for human consumption but, as noted above, reduced for other purposes. Thus farmed products supply about half the global consumption of fresh fishery products, and are therefore valuable contributors to the state of human health and nutrition.

The increase in production since countries first reported production on a regular basis is almost 30%. It is incorrect, however, to interpret this increased numerical production as real growth. The collection of aquacul-

---

Problems of estimating aquaculture production

The greatest problem occurs when fingerlings produced in hatcheries are liberated into inland water-bodies and mix with natural stocks of the same or different species; similarly farmers may produce their own spat of molluscs (mussels and oysters) and grow them out suspended from floating rafts close by to where fishermen may be harvesting natural beds. For convenience sake, data collectors are likely to call the former all natural fish production and record the harvest as inland fisheries, and the latter all cultured mollusc production and record the harvest as aquaculture; sometimes, however, both options are recorded. Another problem occurs when certain warm-water fish are held in captive conditions in different age groups, such as one-year-olds and two-year-olds, before release into open waters and final harvest for market as three-year-olds or more. During the process of collecting data on the weight of fish in reserve each year, the data often are absorbed into statistics for national fish production. Thus the same group of fish may be counted several times over a number of years.

---

ture statistics independent of other fisheries production is, as noted, a recent development. There are now some 147 countries providing annual data to FAO, and the number continues to increase. Furthermore, there are many grey areas of interpretation of aquaculture *vis-à-vis* fisheries practices, and the database is not yet complete.

There are many definitions of aquaculture found in literature, from simply 'fish farming' or 'husbandry of fish and shellfish', to 'the production of aquatic animals and plants above what are produced naturally'. In 1988, FAO defined aquaculture as:

> 'the farming of aquatic organisms, including fish, crustaceans, molluscs, and aquatic plants. Farming implies some form of intervention in the rearing process to enhance production, such as regular stocking, feeding, protection from predators, etc. Farming also implies individual or corporate ownership of the stock being cultivated. For statistical purposes, aquatic organisms which are harvested by an individual or corporate body which has owned them throughout their rearing period contribute to aquaculture, while aquatic organisms which are exploitable by the public as a common property resource, with or without appropriate licences, are the harvest of fisheries.'

There is no single definition of aquaculture which is accepted universally, as there are many forms of human interventions possible in the production of aquatic animals and plants which do not fit precisely. Furthermore, 'production' does not necessarily result in products for human consumption, but commodities for any practical or aesthetic use. It may refer to raising fingerlings or one-year-old animals for stocking or transfer, or for stocks to release into sports fisheries, or valuable species for the tropical fish trade, and even laboratory animals for medical research. A list of aquaculture products and commodities is given in Figure 1.2.

'Aquaculture' is a collective name describing all interventions by man in breeding and/or management of aquatic species. This is analogous to a hypothetical technology for all interventions by man in breeding and/or management of terrestrial species, which might be collectively called 'terraculture'. If such a word was in common use it would describe all forms of agriculture (such as animal, poultry, and crop husbandry), market gardening, horticulture, grassland management, reforestation, beekeeping, breeding programmes for stocking wildlife with animals and birds, horse and domestic pet breeding, and so on. These are all human interventions in response to human needs.

The most important need for aquaculture, like that of agriculture and its related fields, is the production of food for human consumption. Although the most recent food producing sector to emerge in the last century, this relatively modern sector shows much diversity in the ways it is used by mankind, and in the reasons and motivations behind the many approaches. As these are described in the following sections it will become apparent

1. Human Food Products Flesh (or meats) of: • Freshwater fishes, crustaceans, and molluscs • Brackish-water fishes, crustaceans, and molluscs • Marine fishes, crustaceans, and molluscs • Amphibians • Reptiles • Tunicates Roes of fish and reptiles Relishes and sauces Aquatic plants (algae)
2. Live Animals Eggs, embryos (hatchlings), larvae, post-larvae, fry, fingerlings, juveniles, and young adults, of all animal groups in item 1, above Ornamental fishes and molluscs Laboratory animals for medical treatment and research Experimental indicators for bio-assay
3. By-products of Aquatic Animals Processing waste (for meal and fertilizer) Extracts (for medicine and oils) Skins of fish and reptiles (for leather) Nacre of molluscs (for jewellery and decorations) Shells of molluscs (for souvenirs, handicrafts, and extracts)
4. By-products of Aquatic Plants (algae and single-cell animals) Extracts for human food products Food organisms for the culture of aquatic animals Water quality control Experimental indicators and bio-assay organisms Fertilizers Decoration Medicines

Figure 1.2 Products and commodities of aquaculture.

that there is nothing unique about aquaculture. It is yet more variations of animal and crop husbandry, with all their attendant technical, economic, and social complexities.

1.2 The diversity of aquaculture

The fundamental reasons for the diversity of agriculture, and all its related fields noted above, are equally responsible for the diversity of aquaculture. The three principal reasons are:

- the variety of cultured species themselves, each with different life-histories, behaviour, and environmental needs;
- the many forms any intervention by man can take;
- the timing of that intervention in the life-cycle of the species concerned.

Aquaculture development has been assisted by the large number and variety of species which have adapted readily to culture. At present over 150 species of aquatic animals and plants are cultured (see section 1.4.4), and in a range of aquatic environments (see Figure 1.3).

Similarly development has been assisted by many practical interventions by man (see Figure 1.4), noted for his ingenuity and survival, who has husbanded and cultured aquatic animals and plants for his own benefit and pleasure for more than 2000 years with a combination of craft and later technical skill. However, the purpose of intervention by modern man is no longer mere survival, like his forebears, but economic and/or social gains. Consequently, in the twentieth century, the form of any intervention and its timing are directly influenced by the legal, economic, and social milieu in which man now lives.

The evolution of aquaculture has paralleled the history of agriculture, with man moving from a sharer of common-property resources to a private owner of resources, or of rights to those resources, protected by laws of his

Typical areas where aquaculture production occurs:
• Good/low rainfall rain-fed land • Seasonally/flooded land • Partially/fully irrigated land • Inland lakes • Permanent/temporary reservoirs or barrages • Submerged coastal waters • Tidal coastal waters • Submerged coastal substrates • Tidal coastal substrates • Open oceanic waters

Figure 1.3 Production area classes.

The practices of aquaculture, and the measure of production for each:
• Earthen ponds (weight/unit area) • Cages (weight/unit volume) • Raceways (weight/unit flow with time) • Net-pens (weight/unit area or weight/unit volume) • Floating rafts (weight/unit area) • Substrate (weight/unit area, or volumetric measure/unit area) • Trestles or trays (weight/unit area, or volumetric measure/unit area) • Suspended net-bags or lanterns (weight/unit area, or volumetric measure/ unit area) • Open ocean (percentage of release returning and yield) • Hatcheries (throughput per cycle by eggs laid down or yield, or annual production)

Figure 1.4 Aquaculture production practices.

society. Therefore man's interest and investment in modern aquaculture, as with agriculture, are dependent on either his:

- ownership or rights to land, both onshore and offshore,
- ownership or rights to water,
- ownership or equity in operations,
- ownership of livestock,

or any combination thereof. Thus behind a simple concept of 'husbanding aquatic animals and plants', there are now legal involvements of ownership or rights, or equity, for a potential producer to reach his objectives of economic and/or social gains.

Any aquaculture enterprise must be profitable to the investor, although in certain cases 'profit' is not always quantified in financial terms. The most important factors which influence profitability in any enterprise directly concern the need and availability of production inputs, and their cost. Typical production inputs for aquaculture are given in Figure 1.5, but the most important of which are feed, seed, skilled labour, and fertilizer. Other principal costs are availability and price of land, construction of facilities, water, utilities, and local market factors of transportation, infrastructure, and demand. There are also secondary factors which influence economics of production, specifically such things as the general state of technical development (e.g. developments in biotechnology, and information about

In addition to water and land resources, potential requirements for an aquaculture enterprise include:
Seed Live-food organisms for fry/larvae/fingerlings Formulated feeds for fry/larvae/fingerlings Grow-out feed Compost Fertilizer (N, P, K) Pesticides Herbicides Veterinary medicines
Labour (full-time and part-time)
Domestic water Energy utilities (electricity, gas, oil) and communications Mechanical equipment (pumps, compressors) Production facilities (ponds, cages, rafts, hatchery, etc.) Equipment (nets and general farm tools) Support facilities (laboratories, workshop, store, office, etc.) Vehicles (tractor, dredge, plough, trailer) and boats
Permits and technical information

Figure 1.5 Potential inputs for aquaculture production.

it), and availability of supporting services (e.g. equipment manufacturers and suppliers, extension services, professional advisory services, technical training, and education).

Profitability is also dependent on the acceptability of the products of the enterprise, which must be compatible with and acceptable to the established culture of the society as a whole, or its sub-groups. For the most part acceptability is regulated by cultural behaviour, environment, and the economic microcosms of the intended consumers or purchasers of the products (see sections 2.2.3 and 2.3).

Within any society, whether national, regional, or local, the equity or ownership within enterprises by investors, farmers, and, sometimes, hired labourers, and the profitability, influence significantly the strength and direction of the sector. In particular, ownership and profitability direct the decision-making processes, and select the production systems and technical practices most appropriate to make the required return on the investment and labour. This investment may be in the form of capital, but also in the form of time and effort on the part of other members of the society.

Against this complex background, the variability in approaches to aquaculture differ significantly from one economic society to another. Nonetheless, they fit within a relatively simple framework which is described in the following sections.

1.3 The framework of aquaculture activities

A framework of aquaculture development in the last quarter century was first developed and described in Nash (1991). In summary, aquaculture development has been built around three primary groups of activities. These are:

- aquatic farming,
- enhancement of natural fisheries, and
- culture of non-food products.

Within each of these groups there are several sub-divisions which suit the needs of a particular society, community, or type of individual (see Figure 1.6). Where successful, activities within these groups or sub-divisions have been profitable by meeting certain economic criteria, and compatible within respective social and cultural standards. The influence of these factors, and their effects on respective development within the framework, are now described.

1.3.1 Aquatic farming

Aquatic farming is the husbandry of aquatic animals and plants by man. It requires total management of all farm processes, including propagation,

AQUACULTURE PRODUCTION		
Aquatic farming	Enhancement of natural fisheries	Culture of non-food products
(i) Small-scale farming • Owner/operator enterprises • Household enterprises • Nucleus estates (ii) Vertically integrated farming • Owner/non-operator enterprises • Owner/operator enterprises • Incorporated bodies • Collective settlements • Townships • Cooperatives and state farms (iii) Subsistence farming	(i) Culture-based fisheries • Government programmes • Cooperatives and state farms (ii) Recreational fisheries • Government programmes • Owner/operator enterprises (iii) Ranching • Government programmes • Owner/operator enterprises	(i) Owner/operator enterprises

Figure 1.6 The framework of aquaculture activities.

grow-out, harvesting, and marketing of products for human consumption.

There are a number of basic options for the economic production of aquatic animals and plants through aquatic farming which are considered in the decision-making and planning processes. All can be profitable to the operator(s) under certain conditions and circumstances. All these options are widely implemented at the present time. They are not necessarily characteristic of the general level of economy in any one country, and in some countries all are being implemented.

Small-scale farming

Like agriculture, by far the majority of aquaculture enterprises are small scale. Fish farming is predominantly an individual entrepreneurial enterprise for the production of fish and shellfish which the farmer believes he or his wife can sell in local markets, together with any other products raised through integration, such as poultry, pigs, and vegetable crops.

The majority of small-scale farms are classified as *owner-operator enterprises*; that is, they are owned by the producer and/or his immediate family. The owner manages the operations on a daily basis, usually with the help of his wife and extended family, and some permanent or seasonal labour. The financial organization of the farm may be based on a simple company structure, with the farmer and his family owning all shares in the enterprise, or it may be solely on a private basis with the farm providing an income and living for them, with appropriate personal accounting and bookkeeping.

Small-scale farming has many financial and social advantages. Because of the large number of farm units which are responsible for production of a range of aquaculture commodities, the income generation in a region is substantial and, moreover, distributed equitably among a great many individual entrepreneurs. As each is usually a family enterprise, it provides alternative (but usually additional) employment for the women in the family, both in farming operations and in marketing. Futhermore, the operating costs and earnings are within the available capital and cash-flow of any small-scale farmer and his family.

Such farms are usually on land owned by the farmer and/or his family, or on land leased from the government or another private owner. They may also have associated or separate legal rights to water, or may lease water or areas of water (including coastal areas) from the government.

Individual small-scale farms are dependent on the cost and timely supply of reliable inputs, particularly seed, feed, and services. Profitability is also dependent on others in the industry, such as producers of seed, manufacturers of feed and fertilizers, suppliers of equipment, middlemen who process and market the products, and on technical individuals to give advice. In some cases, a number of these external activities, such as the supply of seed and feed, and post-harvest processing and marketing, are coordinated and handled by farmers' associations and cooperatives, but in most cases these are all business activities which farmers negotiate independently.

For technical advice small-scale farmers rely on the assistance of government extension services, service companies, private consultants, other farmers, and sellers of inputs such as fingerlings or feed suppliers.

The size of each farm is usually based on a minimum economic unit for the system, the farming practice, and the species cultured. However, the size may be constrained by many other factors, such as the area of the farmer's holding, the terrain, or the availability of water; or it may be regulated by law for social reasons, as in most of the Scandinavian countries, to fulfil rural economic development objectives, or for environmental and navigational reasons, as in Singapore.

The majority of these farms are indeed small in size. Typically small-scale farms in Africa, for example, consist of one or two earthen ponds, 1–5 ares in total area; in Asia the average size is several ponds, 2–3 ha in total area. There are of course exceptions. Some family-owned and operated farms in Asia, Europe, and Latin America may be as large as 20–50 ha, but are still considered small-scale family-owned enterprises. Many small-scale farms use net pens, floating rafts, and cages. Here, the number of individual units is more important. Typically, a family-owned and operated farm will have at least 2–3 rafts, and 2–10 cages.

The size of each farm is usually based on a minimum economic unit for the system, the farming practice, and the species cultured, or regulated by law

In Egypt the minimum economic unit for brackishwater and freshwater fish which have low value is about 16 feddans (7 ha), and development is scaled accordingly. In other cases the size may also be restricted by law. Regulations in Norway limit coastal salmon farms to 12 000 m^3, and in Finland to 50 t; in Sweden coastal farms are restricted in size and location. These are all for objectives of reducing potential pollution, and developing coastal economies. In Singapore the coastal areas for lease are limited to 5000 m^2, of which no more than 30–40% can be filled with cages. This is also to control pollution and to prevent farms becoming hazards to navigation.

In addition to independent small-scale farming, there are two additional collective approaches often adopted by entrepreneurs and their families. These are:

- household responsibilities, and
- nucleus estates.

Both have the same social advantages of independent small-scale farming in terms of personal income generation and equitable distribution but to a lesser degree, as part of the product and/or profit are repaid to a central organization or the franchise.

Household responsibilities provide opportunities within centrally planned economies for an individual farmer and his immediate family to profit financially through their enterprise. This approach is now widely implemented in China, where members of the old fish farming communes and cooperatives are being allocated ponds for their independent use as small-scale farms. In some cases ownership of the ponds is being transferred to the farmers, with water provided by the state, but in most cases the ponds are leased. At the present time the farmers continue to produce the same crops as their old communes, as these are in demand by local markets.

Nucleus estates, for all intents and purposes, are franchised small-scale farms. An infrastructure, which may be under the control of a private corporate entity, a parastatal authority, or the state, provides a contracted farmer with any modernization necessary, all operating inputs, and buys back the product at a pre-negotiated price. The farmer and his family, or hired labour, produces the required crop, and farm income is commensurate with his success. In Indonesia, the Philippines, and Thailand, for example, part of the shrimp industry is based on a number of nucleus estates with small-scale farmers franchised by larger corporate enterprises. The benefits of this forward-contracting for these small-scale farmers come in the availability of all inputs and the lack of problems associated with marketing the crops. This approach is often used by independent processors as well, who supply all the inputs or arrange credit, and deduct their costs from the purchase price.

Vertically integrated farming

Many large farms may be totally or partially vertically integrated, that is, for economic reasons the owners control all or most of the individual operations which make up the farming processes. They produce their own seed resources, manufacture and supply feed, and many control the processing and marketing of the finished products. Some omit the latter stages, while others include secondary activities such as offering professional advisory services. These individual components may be separate business subsidiaries, or organized within one business under a single or several separate cost centres. The benefits of vertical integration for the producers is control over the timing of inputs and outputs, and the quality and quantity of the products. This has been particularly important for producers supplying international markets, where volume and consistent quality are very important.

Vertically integrated operations are classed as owner/non-operator enterprises or, in the case of communities, owner/operator enterprises. They exist in a variety of forms but all, for the most part, do not have the same benefits in terms of personal income generation and distribution as independent or communal small-scale farming. Most are only alternative sources of employment. This is an important benefit, but vertically integrated farms usually employ only a small core of permanent labourers, with considerable part-time requirements for harvesting and subsequent processing when required. In general, only men are hired as farm labourers, and women as processors. However, the employment for the core personnel is usually full-time, as large vertically integrated farms have the advantage of being better able to shift from one crop to another, either seasonally, or from one year to the next, in response to changing markets and even climatic conditions. Ponds, like fields, are farmed in rotation. Large farms in Hungary and the USA, for example, shift back and forth between aquatic and land crops as market conditions and cropping practices change.

Many large vertically integrated farms are *incorporated bodies*, owned and financed by stockholders. They are directed by a corporate board, and managed by administrators and employees. They may be international in scope, linking joint venturers, or open to all investors. Consequently, these operations have an added level of financial demands, and the return on investment has to be high to meet the corporate structure. As a result, vertically integrated farming is confined to high-value species of fish (such as catfish, salmon, sea-bass, and trout) crustaceans (such as freshwater prawns and marine shrimps) and molluscs (abalone, and oysters).

Because of economies of scale individual farms are large, usually at least 200 ha and possibly more than 1500 ha. Some corporations own many farms, not necessarily all in one country. The core of permanent employees may include managers, administrators, and general service personnel, trained in their respective fields but not necessarily with respect to aquaculture technology; technical managers, scientists and technologists, technicians and field staff (labourers for pond maintenance, feeding, harvesting, etc.), and engineers and mechanics. In addition, there may be processing plant managers and processors, and feed manufacturing plant managers and operators, both with engineers and mechanics. The majority of these positions are occupied by men, who have more opportunities for education and technical training than women, but many trained women are gainfully employed in hatcheries and broodstock/nursery farms, where they have proved to be better technicians.

Vertically integrated farms privately owned by incorporated bodies are usually associated with developed countries, but this is not always the case. Both international and national corporate enterprises control many of the large farms for the production of marine shrimp in Ecuador, India, Indonesia, and the Philippines, and for the production of salmon in Chile. Most export their products to markets in Europe, Japan, and North America.

A number of *collective settlements* in both developed and developing countries include fish farming among their many enterprises. In those settlements where there are suitable opportunities, the culture of marketable fish and crustaceans is widely practised, together with many other production-oriented industries. The settlements are profit-oriented, and fish are produced for national markets as well as for the needs of the settlement itself. The fish farming operations invariably include all components of production and marketing.

The most advanced level of collective settlements are the kibbutzim of Israel. These, which are all owned and operated by members, are not confined to food production but include a wide range of light manufacturing industries and services. Many undertake their own applied research to support production. Fish farming is a common activity within the kibbutzim, and invariably includes all components of production (hatchery operations and grow-out), as well as processing and marketing.

There are direct economic advantages within the collective settlement which assist profitability of fish production, as well as profitability of the settlement as a whole. In particular there is the availability and use of communal labour, particularly women, trained on site to support the small number of (full-time) trained technicians, and the close integration with other settlement activities and resources, such as the use of organic wastes from animal and poultry units for fertilizer, refuse and kitchen waste for feed, and common vehicles for transportation and marketing of products.

Collectives have many external social advantages in terms of benefits to the farmer and his extended family. There is access to other social services generated by the community, the availability of other cheap food commodities which are produced by the group, medical services, and the generation of other family employment in the larger structure.

In summary, collective settlement farms are productive and economic to operate. They play an important part in the national food production sector, and contribute significantly to the national objective of food security.

Integrated farming is the communal production system prevailing in China at the *township level*. These production units, which are owned by the community collectively on land owned by the state, invariably have a single industrial focus, such as animal husbandry, tea, fruits, or forest products. However, just like the kibbutzim of Israel, many townships enhance their profitability or provision themselves with food by integrating with other activities. For example, a typical fish farming township in China may grow mulberry for rearing silkworms, raise sugar cane, and grow vegetables, all in and around the fish ponds.

The labour force in a communal township is large, as it includes everyone to some degree. It is divided into brigades and teams of men and women, made up of both skilled individuals, educated and trained in the technology, and unskilled helpers who may, however, be skilled in other crafts. Each brigade and team is responsible for specific daily tasks. Members of the commune may work in several different teams (for technical and domestic tasks) and in different brigades through the year with respect to seasonal demands of crops.

Township level farming, most common in China and certain other countries of Asia, has developed from traditional communes. Although most of these townships derive their incomes from the sale of food and agriculture products, notably fish, pigs, and ducks for local markets, a number now joint-venture with overseas companies for the production of high-value marine species for export, such as shrimps, and are also providing technical services.

A number of countries, such as Cuba, Hungary, and those in Eastern Europe, have (or used to have) an aquaculture sector based entirely on *cooperatives and state farms*. A cooperative may be a fixed enterprise on contiguous land sub-divided into individual farms, or a flexible enterprise for the support of individual farmers on their own farms spread over a large area. Cooperatives, which are owned by their members who have collective or individual rights to the land, are invariably self-sufficient in providing all the required farm inputs, and are relatively independent. Some cooperatives are not unlike collective settlements and have several production-oriented businesses and extramural activities (such as retail marketing and even restaurants) all of which contribute to the overall economy of the cooperative, and generate employment for all members, both men and women. In large agriculture cooperatives the fish culture activities may be quite small.

State farms, on the other hand, are owned and operated by the state. There is no individual ownership by farm managers and possibly hired labourers, all of whom are government employees, and usually only men. Land and water are owned by the state, although there may be inter-ministerial legal and financial settlements. For state farms specifically for aquaculture, the control is under the respective government department. The state system organizes and manages all associated farming activities, such as seed production, and marketing.

In Hungary, for example, the aquaculture sector was, until recently, entirely operated by cooperatives and state farms. Most of the cooperatives are agriculture based, with fishing and fish culture as supplementary activities. The state farms, in addition to production, supply fish for stocking in inland waterbodies and for sports fishing. As might be expected, the manpower used in the management and operation of state farms describes a great range of education and technical skill. Training is provided within the system, as needs require it, and individuals invariably have many responsibilities.

Subsistence farming

The third organizational and management option for aquatic farming is subsistence farming. It is carried out by individuals and groups of individuals who farm aquatic animals and plants on a part-time basis as one of several other part-time agricultural activities. It is attractive for a number of reasons, but the most important are probably:

- the use of available land and water resources which cannot readily be used for other production,
- increasing the variety of foods available, especially increasing the value of fish as a 'relish' to enhance the palatability of staple foods, which is probably more important than the value of fish as animal protein,
- gaining status in the social community, by owning a pond as well as producing an unusual crop which is especially valued as a gift in many subsistence level communities.

Subsistence farms produce low-value food fish to provide animal protein for the rural farmer and his immediate family, with any surplus bartered or sold to the village community. Thus subsistence farming generates little and often nothing in the way of personal income for the farmer or his immediate family and, in spite of the large numbers of people employed, employment is mostly part-time. In some countries daily operation of a subsistence farm is the customary responsibility of women, adding to their already heavy work-load, and in others it is that of the men.

Subsistence farming is characteristic of rural development projects organized by governments as part of a national agricultural development programme. Such programmes try to provide large numbers of rural inhabitants with a small fish production unit to manage in addition to several other agricultural production units producing (say) cereal crops and vegetables. In some cases the participants will own the land or be given rights to the land. Invariably none of these basic fish farms is independently economically viable (as they are small in size and often only rain-fed), and their maximum production potential does not meet minimum nutritional requirements for the farmer and his immediate family. Together with other production activities of fruits, cereals, and vegetables, however, and household goats, pigs, and poultry, they make an 'economic and nutritional whole'. These agro-aquaculture systems are also economically viable for small semi-subsistence farmers where suitable land and water are available, and employment opportunities and nutrition are critical needs.

For the production of low-value food fish the type of farming practised is extensive. The ponds may be very small (1–2 ares in area) and shallow. Almost no inputs are required. The fish, if managed correctly, maintain their own numbers, and feed on the natural productivity of the ponds supplemented with household vegetable wastes and composts.

Subsistence farming for freshwater fish is characteristic of many African countries for the production of native tilapia species. This is a hardy group of fish which tolerates a wide range of environmental parameters, and survives and breeds in poor conditions often associated with part-time farming. Few shellfish are adaptable to subsistence farming, although poor landless communities in Chile work mussel beds together with their primary activity of gathering edible seaweeds to earn income. The majority of subsistence farmers have little or no background education, and rely almost entirely for technical direction on the government extension service.

Although subsistence farming is often proposed as a food producing activity for landless labourers, it proves to be difficult in practice. The need for daily management and continuous surveillance is often contrary to the lifestyle of such people, and ponds are usually abandoned once the fish have been harvested or lost.

1.3.2 The enhancement of natural fisheries

A considerable part of aquaculture world-wide is carried out to enhance or directly support natural fisheries. Many aquatic species are cultured in captivity and released into the natural environment for subsequent harvesting by fishermen. Thus breeding adults or fingerlings from hatcheries are used traditionally to stock large inland water-bodies, reservoirs, and permanent and temporary water barrages, and spat of juvenile molluscs are also seeded out on natural beds or fixed structures in coastal waters. In recent years the practices of stocking and release have been extended to both fish and crustaceans in open waters.

The two principal objectives of culture-based fisheries are:

- to create or enhance a commercial fishing industry, open to all, and

- to ranch animals and plants in open waters to avoid the costly inputs of constructing and operating facilities and providing feed.

The enhancement of natural fisheries has little social benefit to those involved in creating them. They are mostly publicly managed and operated, and require few but highly-trained employees. On the other hand, they have considerable benefit to large numbers of fishermen. They are particularly useful to artisans, who operate small-scale family-type fishing operations which are labour-intensive and require low levels of capitalization, technology, and productivity; but they can also benefit the industrialized fishermen, who operate commercial capital-intensive vessels requiring shore-based infrastructure consisting of harbours, repair and maintenance, cold storage, processing, and marketing facilities. Thus enhanced fisheries have a far-ranging effect on employment and income generation and, compared with aquatic farming, produce these beneficial effects within a far shorter period of time.

Culture-based fisheries

Many commercial fisheries of large inland lakes, reservoirs, or in compact coastal zones, are culture-based. Natural populations are enhanced by the release of breeding adults, or stocked with fingerlings of fish, or spat of molluscs, such as oysters and abalones. The resulting population is therefore freely available to the fishing community under any existing management regulations, and free to sell on local markets in the traditional way.

As these are communal fisheries, with no ownership rights (except by the state), resources for culture-based fisheries are supplied by public hatcheries and state farms. The economics of operating these facilities are not always analysed as independent cost centres but in terms of value of the end fishery, and the industry itself within the national economy. Therefore publicly-owned and operated hatcheries are not always managed and operated efficiently. The financial returns to the government come directly in the form of licences or levies, and indirectly in the form of a national industry.

Where fisheries are identifiable and owned, the hatcheries are also owned by private producers. The cost of operating private hatcheries is built into the overall economics of the individual businesses. Many private producers form associations or cooperatives for harvesting, processing, and marketing activities, and for leasing appropriate space for production in communal but protected waters.

Harvesting culture-based fisheries may be by large commercial vessels or coastal traps, or more typically by the local lake communities in small fishing boats and canoes. In small water-bodies, such as the irrigation 'tanks' of India and Sri Lanka, or in the water barrages of Africa, villagers use traps, or fish directly from the shore. Traps are also used to recapture lobsters released as part of a national enhancement programme in the

USA. In the coastal fisheries for oysters and mussels in France and Spain, which are owned by private producers, the spat are produced in private hatcheries.

Recreational fisheries

Government enhancement programmes for recreational fisheries, using cultured fish, are another valuable contribution of aquaculture. Recreational fisheries are, for the most part, owned and operated by the state, and managed through licences and regulations. Many recreational fisheries are more an available resource of food for the fisherman and his family, and not merely sport.

The production of fish for release into recreational waters is carried out in national and state hatcheries. In association with enhancement the administrators of the programme operate a management and policing function. The return on investment to the government comes in the form of licence fees from both local sports fishermen and tourists, and the national micro-economy of fishing equipment and secondary industries (boats, accommodation, food, and fuel sales). These all can be considerable. In Hungary, for example, there are over 300 000 registered anglers who fish inland waters stocked by the government through its cooperatives and state farms to increase the availability of food fish for the public.

Many privately-owned ponds and streams are also popular for recreational fisheries, whether in the form of farm ponds managed for recreational fishing, or of the 'put-and-take' type, where the operator charges for the right to fish and the fish caught. These private enterprises are common in the USA and countries of Europe.

Enhancement through ranching

Ranching is a relatively new variation of enhancement, based on aquaculture. Ranching is usually confined to high-value migratory species which return to the point source of origin, and which can be reared cheaply in large numbers to compensate for subsequent losses in the wild. The most common examples are the Pacific and Atlantic salmon, which return to their points of release. Canada, Chile, Japan, and the USA all have public or private ranching programmes for salmon. It is known that all these releases enhance the commercial resources offshore, but enough return to the point of origin to provide a return on investment for the producers. But there are other species with potential, such as eels, and some crustaceans and molluscs which will stay within certain locations. Japan is developing ranching programmes with other non-migratory species of fish (such as seabreams), crustaceans (lobsters and marine shrimps), and molluscs (abalones), and attempting to ranch species by adapting their behaviour to retain them within certain coastal areas for subsequent harvest by those who release them.

With respect to manpower resources, all three enhancement activities described above (culture-based fisheries, recreational fisheries, and ranching) require cadres of well-educated and technically skilled individuals to manage the programmes and produce the stocks for release in hatcheries. The beneficiaries are national fishermen, anglers, and in some cases private individuals. They also require educated and trained individuals to police the programmes and collect licence fees, and compile statistics for management purposes.

1.3.3 The culture of non-food products

Aquaculture includes the production of a variety of non-food products (see Figure 1.2). It contributes significantly to the *ornamental and tropical fish industry*, and provides an alternative source to natural stocks, thus helping to mitigate environmental damage caused by harvesting natural resources from offshore tropical reefs. It also supplies the *bait-fish industry*, which is equally important in support of recreational fisheries. Aquaculture is a supplier of *luxury commodities*, such as cultured pearls (from oysters) and leather goods (from skins of crocodiles, salmon, and eels), and shell-crafts (from turtles and giant clams). It also produces *live food organisms* for other cultured species and the tropical fish trade, and laboratory species for *medical research, medicines and medical aids*.

These are all activities vested in the private sector, and their manpower requirements are small in number but mostly for highly-trained and specialized individuals. Although the culture of non-food products is an economically important part of the aquaculture sub-sector, the benefits of income generation and distribution are very small compared with aquatic farming and enhancement of natural fisheries.

1.4 Factors which influence the direction of the sector

The previous section used a simple framework to categorize the management interventions of man to meet his specific economic, social, and cultural backgrounds. That being said, however, the selection of the most appropriate direction to suit these interventions is based on many physical, chemical, and biological factors, such as the terrain, the quality and quantity of water available, and the suitability of the aquatic animals and plants which are to be maintained. Just as land-based farming has been adapted for innumerable animals and crops in differing climates and specific locations, aquatic farming is rapidly evolving the same way. Consequently there are a large number of systems and practices now available to modern aquaculture technology, and many species to choose from. These are described in the following paragraphs.

Aquaculture is invariably compared with traditional capture fisheries because both industries produce the same marketable commodities.

However, the aquaculture sector is more related to agriculture than to capture fisheries, and the principles and systems of agriculture help to identify and accept the principles and systems of aquaculture.

In spite of the number of species and diverse range of aquatic animals and plants currently under production, modern aquaculture, like agriculture, can be categorized broadly into three basic systems based on the intensity of production. These are:

1 *extensive aquaculture*, for organisms cultured in low densities, dependent on natural productivity for food but possibly assisted by fertilization of the substrate;
2 *semi-intensive production*, for cultured organisms at higher densities (than would be the case in extensive culture) and dependent on both increased productivity, using fertilizers and waste organic resources and kitchen refuse, and receiving supplemental artificial feed; and
3 *intensive production*, for cultured organisms in high densities, and dependent on artificial feed for their nutritional requirements.

The principles of these three fundamental systems are simply defined, but the biological and engineering techniques which are used to apply them are manifold. Furthermore, the techniques frequently overlap each other, and different techniques are often used for different phases of the production cycle of a single species; for example, juveniles may be reared intensively in tanks, but grow-out is extensive in ponds. Consequently there is a vast matrix of options and alternatives for the potential producer to consider, and all carried out in a three dimensional medium (water) which, as noted, can be fresh, marine, or something in between.

Their characteristics are briefly summarized here. In particular they emphasize an increasing level of risk to the producer, hence the requirement for differing levels of technical skill to minimize that risk.

1.4.1 Extensive systems

Extensive systems utilize low-level technology requiring few inputs. Hence they are predominantly characteristic of aquatic farming at the subsistence or semisubsistence levels (see section 1.3.1), but also, by definition, they are a feature of enhancement (see section 1.3.2).

One man, with the help of his wife and extended family, is invariably capable of managing the majority of subsistence and semi-subsistence farms which adopt extensive practices, with the help of trained extension agents to provide technical advice and guidance. Extensive farming has changed little through the centuries and farms are exposed to all the vagaries of floods and drought, or predation and theft. The only species ideally suited for the extensive system are those which require little attention, breed prolifically, and survive a wide range of natural conditions.

Yields from extensive production are low

Typical yields of crustaceans in the extensive system, such as crayfish in rice paddies, are about 500–1500 kg/ha annually, but with continuous cropping (selective daily harvesting of individuals above a chosen size). Annual yield of low-value fish, such as tilapia species, may be no more than 400–1500 kg/ha in small earthen ponds.

The principal purpose of extensive farming is low-cost production of food for the farmer and his family, with any surplus for bartering or selling. However, the continuous demands on small units and harvesting of undersized individuals makes typical yields hardly enough. Fortunately, the investment cost is little more than the time of the farmer to construct a unit, perhaps with locally available materials, and to seed it with some breeding adults. Moderate increases in yield can be obtained by increasing water productivity, and by careful attention to culling and harvesting techniques.

The benefit of extensive farming is the improved nutrition of the farmer and his family

In a study in Guatemala improved nutrition was the primary motivating factor for most of the 8000 family members participating in a development project. Average fish production from a median size 120 m^2 pond was sufficient to increase dietary protein intake of a typical seven-member family by 28%. This high quality fish protein further enhanced the nutritional status of rural families by supplementing deficiencies of low quality protein in their traditional diets.
Source: Castillo *et al.*, 1992.

The most common practice of extensive farming is production in individual ponds, regular or irregular in shape, which may be rain-fed or supplied with running water from some source. Mostly low-value finfish, such as tilapias, are raised in such ponds, surviving on pond productivity for their food, perhaps supplemented with household wastes, including wastewaters, and vegetable composts. The annual production in fertilized ponds without supplemental feeding is usually very low, and a good year might yield the farmer about 500 kg/ha (equivalent) of fish of differing quality, quantity, and value. At the semi-subsistence level, the financial return from several ponds may be sufficient for a maintenance income for the farm and farmer, but insufficient for expansion or the regular repayment of debt, or purchase of feed and fertilizer.

Women and children participate actively in extensive fish-pond production in rural areas, particularly in Central America and West Africa, possibly accounting for half the total labour required. Among the recorded labours of women in pond production are brush clearing, digging, transporting earth, construction, pond cleaning, stocking, feeding, fertilizing, raising ducks and pigs, harvesting, processing, and marketing.

One alternative to the prepared pond is the use of existing rice paddies, and fish and certain freshwater crustaceans, such as prawns and crayfish, are produced extensively with rice. Although this type of farming is seasonal, it is well suited to species which require little husbandry, and fish-cum-rice farming, as it is called, is widely practised in irrigated rice paddies in rural areas to supplement family nutrition.

Included in extensive systems are practices which rely on the ranching or free-ranging of cultured species in large water-bodies, so that there is no investment in feeding and labour. This is 'enhancement', which has been described in section 1.3.2, above. These water-bodies can be both closed (such as a lake or reservoir), or open, such as an inland sea or open sea. However, the enhancement of such fisheries is culture based, that is they require the intensive production of juveniles by a few well-trained technicians, usually in a publicly-owned hatchery and nursery.

Enhancement of fisheries through aquaculture benefits large numbers of artisanal fishermen

A small carp hatchery, producing about 2 million hatchlings (four-day-olds) and 1.5 million fingerlings (25–50 mm in length), only requires one skilled manager (or owner), a trained technician, and two competent labourers. This should provide at least 2000 t of fish in a local fishery for the livelihoods of many small-scale fishermen.

The same principles of avoiding feed and labour costs also apply to the practice of free-range farming of mussels. But, as mussels are sessile organisms, they are suspended on ropes beneath floating structures. No hatchery is required, and the practice is well-suited to many families in coastal communities. Although the capital investment in a sturdy raft is high, the yields are considerable if the productivity of the water is high.

Many coastal communities practise ranching of molluscs and obtain good yields

In France and Spain, the annual yield of a 500-rope floating raft is about 50 t of mussels, and requires one person to manage it; thus each family owns on average 2–3 rafts. Larger operators may have up to 25 rafts and employ 10 people.

1.4.2 Semi-intensive systems

Semi-intensive systems are the most common, as farmers try to increase production by increasing husbandry and farm management, manipulating water flow (to provide oxygen and maintain water quality), and supplementing natural foods with artificial feed. They have higher risks than extensive systems and higher investment, but the increased yields provide good profits in well-run enterprises. Therefore they are the basis of all aquatic farming (see section 1.3.1), and the backbone of entrepreneurial small-scale farming as well as vertically-integrated farming.

Yields from semi-intensive systems are high

Typical production levels are 750–1500 kg/ha per cycle (and farms in tropical countries obtain 2–3 cycles per year) for shrimps and prawns, and 350–400 t per annum of three-year-old oysters on about 100 ha of tidal zones (including the areas for first and second-year oysters). Annual yield of fish in ponds can be between 2000–5000 kg/ha.

The popularity of semi-intensive farming is due to both technical progress and, in some cases, the retreat of intensive operators who have found risks at that level too great (see section 1.4.3). All semi-intensive systems require many inputs (see Figure 1.5), and therefore the owners/operators require investment capital for construction, annual supplies of seed for stocking, and quantities of suitable feed to supplement the productivity induced by compounded fertilizers. The availability of capital therefore dictates their level of operation.

Most small-scale farmers attempt to increase the level of intensity as much as possible, but economically trading off between survival and production, the cost of feed, and the cost of labour. They also prefer to produce crops which have the highest market value, some of which may be possibly out of the reach of local markets. Consequently, as aquaculture technologies for high-value species have advanced in the last decade, more fish farmers are converting their farms to the production of crustaceans (shrimps and prawns) – particularly in countries in Asia where fish farming has been a tradition and many families already own suitable ponds. Therefore semi-intensive systems are adopted by producers with the intention of increasing family income, with secondary benefits of family nutrition and barter.

Small-scale farmers use semi-intensive systems to increase family income

In Guatemala on an average farm the fish pond occupied only 2% of the farmer's land, but the market value of all fish harvested increased on-farm net income by 18%. Expressed in terms of employment potential, a typical pond required 0.7 man-months of labour annually, but net cash value of the fish crop was equivalent to approximately two months of wages as a rural labourer.
Source: Castillo *et al.*, 1992.

The cheapest way for farmers to increase yields is by increasing natural water productivity through additions of nutrients. These may be in the form of inorganic fertilizers, organic composts, dry and wet waste products from other animal and poultry production units, and even human sewage. Increased water productivity is an important supplement of the diet, as the natural food organisms encouraged to grow within or around the production units supply the stock with specific amino-acids, vitamins, and trace elements missing from artificial feeds.

The benefits of the multiple resources of a communal township in China (Lelai People's Commune in Guandong Province)

Commune Background	
Surface cultivated	4 950 ha
Population	
Number of persons	91 000
Number of families	17 000
Production brigades	29
Production teams	235
Integrated agricultural:	fish, sugarcane, mulberry production, trees, silkworms, livestock (mainly pigs)
Industrial production:	glass, paper, printing, plastics, alcohol, oil, tools, medicines, electronics, boats, bricks, limestone, agricultural machinery, insecticides
Annual average income *per caput*	Yuan 260
Fish Culture	
Production surface	2 400 ha (8350 ponds)
Annual production	7 700 t food fish 140 million fry
Percentage total income	30%
Cultural system	Semi-intensive polyculture

Source: ADCP/REP/88/31, FAO, Rome.

Increased production through integration with agriculture (particularly the husbandry of land animals, poultry, and even certain cereal and vegetable crops) can be an advantage for both large- and small-scale investors. Because of the need for close management of highly interdependent operations, enclosures are often small. Such integrated systems are not widely used for non-fish species. Public health standards require molluscs exposed to environments rich in microbial organisms to be depurated before marketing, and such environments are not always conducive to the survival of crustaceans. However, freshwater prawns are often cultured in nutrient-enriched ponds in Asia, together with fish.

Nowhere has integration been more skilfully accomplished than in the collective settlements of Israel and the communal townships of China (see section 1.3.1) where most fish are farmed with other livestock and crops. This type of farming is labour intensive, but there are manpower resources and other communal assets on which to draw to make this type of farming very efficient and productive.

The semi-intensive system is typically based on practices which use large regularly-shaped ponds and depend on pumped water. Ponds are used for both freshwater and marine fish and crustaceans. Pond construction and the energy required to pump water is often costly and farmers try to minimize the cost by using family labour and reduce the need for water exchange.

Production in earthen ponds is the most popular semi-intensive production practice

A desirable target for small-scale farmers using fertilizer, minimal supplemental feed, and minimal water flow from any water system, is about 1000–1500 kg/ha (equivalent) of fish, and possibly an additional 500 kg/ha of shrimps or prawns if fry resources are available. The cost of construction is cheap, depending on size, as most small-scale farmers rely on family labour. In the Guatemala project noted, for example, family labour, a non-cash cost, accounted for 84–91% of total pond construction cost.

Alternatives to ponds are cheaply-constructed net-pen structures which enclose a body of water and also avoid the need to pump water. These also have few restrictions on size. The same advantages apply to farming fish, crustaceans, and molluscs in permanent or temporary barrages. Inland barrages are being used in many countries for the farming of freshwater fish and prawns, and coastal barrages for fish and molluscs. Barrages have proved to be useful communal farms requiring minimal attention, other than stocking of fingerlings and regulated harvesting.

Nonetheless, small-scale family-owned units with little access to credit for expansion can only intensify so far, and cannot compete with the larger commercial corporate investors, particularly those operating vertically-integrated farms. Consequently in Indonesia, for example, there has been 'dualistic development', with modern highly capitalized almost intensive systems existing side by side with the traditional low-input system.

Compared with small-scale farms, vertically-integrated enterprises are capital intensive and require more hired labour

In Indonesia, modern semi-intensive vertically-integrated shrimp production farms require about 14–30 hired workers (usually only men), and the traditional system between 1–3 family workers, per 10 ha of ponds. A small-scale 2 ha family-owned shrimp pond operation, which may produce 0.5–2 t annually in two crops, provides a significant income for the family, and requires only 30 days of family labour and 60 days of hired labour each year. This compares with 32 and 120 days, respectively, for rice cultivation in the same area. Modern farms produce 10–15 t/ha through multiple cropping, and require full-time attention.
Sources: Bailey, 1992; Hannig, 1988.

The situation in Indonesia is not necessarily characteristic of other countries, where there has not been the intense national investment in nucleus estates (see section 1.3.1) sustained by international loans. For example, the operation of a 20 ha semi-intensive shrimp pond of a large-scale operator in Panama requires one full-time unskilled labourer and one full-time pond manager, with an additional 10 days of labour during harvest. Thus many semi-intensive farms in Central America are still family-owned and operated independently.

Because shrimp production is highly technical and a valuable contributor to foreign exchange in many Asian countries, administrative and management personnel are professionally educated, particularly in the biological fields. Educated and technically-trained individuals are employed full-time as technicians in the hatcheries, nurseries, grow-out units, and speciality areas such as feed production, disease monitoring, and engineering. There are also large numbers of full-time unskilled labourers on each farm for routine work such as feeding, sampling, stocking and harvesting, maintenance, and mechanical work. Part-time staff are often hired at harvesting.

1.4.3 Intensive systems

Intensive systems are the most productive per unit area of farm. As the principal purpose of the aquatic farmer is to make a financial return on his investment through the sale of marketable commodities which he produces, the direction of the technology has been towards advanced practices based on the intensive system. Investment and operational costs of an intensive system are high, and so is the risk, and therefore management and operational efficiencies must be constantly monitored.

For profitability, the practices of intensive systems require large yields per unit volume of water (in terms of space or flow). The yield may be one species or several cohabiting species, which is often possible in well-managed natural environments. Most practices are highly technical, requiring properly engineered construction, mechanical operations, monitoring of water quality, and preparation and provision of nutritionally acceptable feeds. As the systems are capital intensive, intensive farming is predominantly carried out by corporate entities operating vertically-integrated operations. However, in developed countries there are a few independent small-scale operators producing salmonids intensively where natural water conditions are ideal.

Intensive systems are the most productive per unit area/volume of farm

Typical intensive systems may carry 30 000–90 000 kg of mussels (wet weight) on floating rafts (300 m^2 in area), or 2–5 kg/m^2 of shrimps in raceways. Intensive production may yield annually from 5000 kg/ha of catfish in land-based ponds, to 10 000 kg/ha equivalent in net pens. Production of trout is typically measured in weight per unit flow with time, or weight per unit volume. A litre of water per minute can sustain about 2.4 kg of rainbow trout in intensive conditions, increasing to over 9 kg with added strong aeration; alternatively, densities of about 15 kg/m^3 can be maintained in offshore floating cage units, which may be from 32 m^3 to 30 000 m^3 in volume.

Because of economic advantages of the intensive system, specifically large volumes of product from small areas of water surface and land (which may be submerged or tidal), there are many options for the producer, all of which are appropriate technology. At an intensive level earthen ponds are elongated into man-made channels, but the most common production units are concrete raceways or tanks through which large volumes of water are delivered to provide oxygen and remove waste products. One way to avoid the problem and cost of water delivery is to contain aquatic animals within naturally-occurring water-bodies, including the open sea. Consequently many finfish are cultured in constructed cages or net-pen enclosures which rely on the natural movement of the water to maintain water quality, but require high-protein feeds. Molluscs, such as oysters and clams, are also raised intensively in hatcheries and grown-out in open waters, but on trestles or trays, suspended from rafts on ropes or in bags, or sown on specially prepared beds on the substrate.

Because of the high risk of intensive farming it is necessary to have a good understanding of the biological processes of the cultured species. Consequently the manpower required to manage production practices which are intensive is invariably small in number (to minimize costs) but well-qualified and highly trained in a range of specialities, such as fisheries biology, engineering, nutrition, and possibly pathology. In many cases this has to be the farmer himself, as the minimum size of a trout farm, for example, is between 12–20 t per annum to provide a living for one man without any other income. Consequently, many intensive but small-scale family-owned farms survive, but often at the expense of subsidized or non-costed labour from the family.

Hatcheries also are considered to be intensive production units, and the same characteristics apply. Public hatcheries are usually over-staffed, but private hatcheries reduce labour to a minimum. Small family-operated 'backyard' hatcheries have proved to be highly productive and profitable, and much of the shrimp and prawn industry in Thailand and Taiwan has grown on the strength of these enterprises (and to a lesser extent in Malaysia and Vietnam). But in Indonesia and the Philippines, and also in Ecuador and Panama, the industry has been planned to grow through private investment in large commercial hatcheries in support of nucleus estates. Hatchery work is well suited to women, which is probably why most backyard hatcheries have been so successful; but many women are employed in hatcheries throughout the world.

In conclusion, it is apparent that all three production systems and their multitude of technical practices are options for the small-scale farmer or the corporate investor in the private sector to consider, as well as the government administrator. Essentially they all make the same demands on major resources, such as capital, labour, land, water, food, etc., but obviously to greatly differing degrees. Similarly they fulfil different objectives.

Therefore it is not surprising that a national aquaculture sub-sector, considered in its entirety, can consist of all or any of the possible combinations, as each one may be economically or socially justifiable in its own right, depending on the availability of major resources, noted above. However, aquaculture has to compete for these resources with other sectors, and some of the most conflicting issues are briefly described in Chapter 2 (section 2.1.1).

1.4.4 Selection of species

The principal criterion for the selection of species for use in each system and practice is market demand. The growth in demand of the principal aquaculture commodities is illustrated in Figure 1.7.

Specific species suit certain markets and therefore the choice is important. Growth rates of cultured species are also important, and they are higher in the tropics than in temperate climates due to elevated water temperatures. For the most part, tropical and sub-tropical countries produce

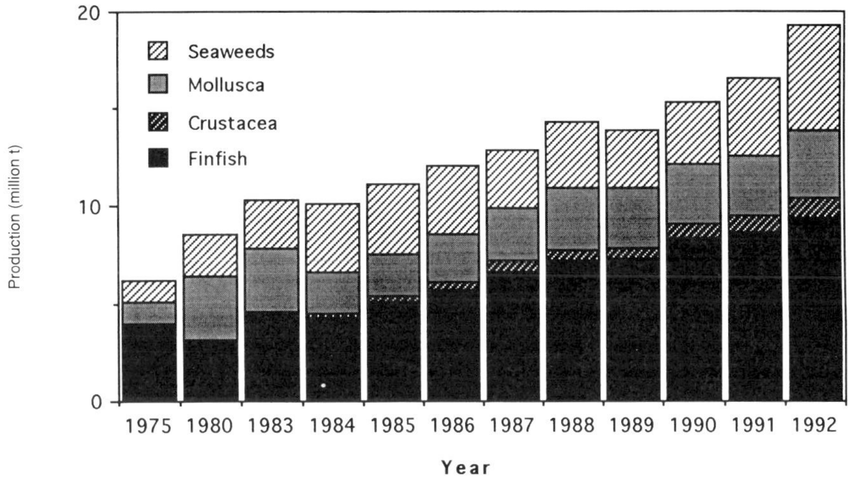

Figure 1.7 Growth in production of principal categories of aquatic animals and plants since 1975 (in million t) (Source: FAO, 1994).

a far greater range of species than temperate countries. Temperate countries invariably confine culture to high-value marine species, because of the high level of investment and the regional markets.

As a general rule herbivorous fish are the cheapest to produce; their value is low but they can be produced in large quantities. For the most part all brackishwater and marine fish, and all crustaceans, are carnivorous or omnivorous; they are more costly to produce as they require high-protein feed, but they have a high value. The majority of molluscs, which are filter feeders, are moderately costly to produce and have a range of values; on

---

There are 155 identified species of aquatic animals and plants under some form of culture

Of the 'species' currently cultured, 89 are finfish, 23 crustaceans, 35 molluscs, 4 seaweeds, and 4 others. Production is dominated by 19 species of carps and related fish yielding almost 7 million t, followed by 9 species of oysters (about 1 million t), and 10 species of mussels (over 1 million t). Eight species of tilapias and cichlids (about 500 000 t) were followed by 11 species of salmonids (over 600 000 t). 26 species of miscellaneous freshwater fish, such as catfish, perch, and snakehead, make up a further 1.5 million t). There are also 11 species of clams (about 800 000 t), and over 17 species of shrimps and prawns (about 900 000 t).
Source: FAO, 1994.

---

the other hand they can be produced easily in large quantities. Seaweeds require little more than a suitable natural environment for culture, and their value is mixed depending on whether they are produced for human consumption or for industrial use.

Within the major categories of species, large numbers of individuals are currently cultured.

1.5 Forecasting the direction of the sector

Almost all national governments include food security high on their list of priorities for planning and investment, and most countries have some technology in place to enable aquaculture to be included as one component of any national plan to achieve their objectives. However, the planned direction of any food-producing sector, and its strength, must consider national, regional, and even local trends in several factors, all of which are continuously in a dynamic state.

For the aquaculture sector, key factors are the availability, price, and consumption of fisheries products and directly competing products, such as poultry and white meats (pork and veal). Also important are key resource materials (such as fish meal, which is the animal protein base of all animal, poultry, and fish feeds). Then there are the traditional indicators of demographic changes (population growth, migration, and labour availability), and of course resources of energy, water, and land, etc., followed by social patterns.

The majority of fisheries products come from capture fisheries. The forecast for the future of capture fisheries landings, which is made in global terms by organizations such as FAO, together with the opinions of leading fisheries scientists, is continuation of small increments (less than 1% per annum) attaining a potential of about 100 million t from marine fisheries, and about 20 million t from inland fisheries (excluding aquaculture). Some individual forecasts put marine fisheries as high as 120 million t due to the harvest of new or underutilized species, and improved ways of handling the by-catch. Increased consumption of the products may be obtained by post-harvest technology, but this will not affect the total harvest.

The demand for fisheries products continues to increase

The increase in demand for fish for direct human consumption is forecast as an additional 30 million t by the year 2000. This might be satisfied by better fisheries management (10 million t), improved utilization of resources (20 million t), and possible increases from aquaculture (about 5–10 million t). Supply constraints are likely to increase prices which will improve the viability of aquaculture, but, at the same time, constrain overall demand.
Source: FAO, 1981.

The same forecasters indicate that the major potential for growth in the production of fisheries commodities, and particularly fresh fish, is therefore through aquaculture. Current projections indicate that countries currently active in aquaculture collectively will be producing about 22 million t of fisheries commodities and aquatic plants by the year 2000. Most of this production will be the same fresh products which are already contributing significantly to the percentage of fresh products in the range of commodities from the annual global fisheries harvest (see Figure 1.8).

Thus aquaculture will account for approximately 25% of world fisheries production, and probably half the value. The projection for the growth of global aquaculture production is therefore an average of over 5% annually. This is a reasonable projection, given the emerging state of the sector and its current high profile, but it should be noted that the distribution is unlikely to be uniform throughout the four major commodities (finfish, crustaceans, molluscs, and seaweeds), or throughout the five main geographic regions (Africa, Asia/Pacific, Europe/Near East, North America, and South America). Many countries have yet to establish the critical mass of producers within the sector to achieve this growth rate.

The demand for fish and fisheries products, from whatever source, will continue with the increase in global population, which also differs from region to region and country to country, The projected average annual growth rate of the world population to the end of the century is 1.7% (range 0.4–3.4%). The world population in 1992 was about 5000 million. Fish in particular offers a realistic alternative as one of the cheapest sources of animal protein, competing with poultry and cheap white meats.

Fisheries commodities are also responsive to current social trends. For

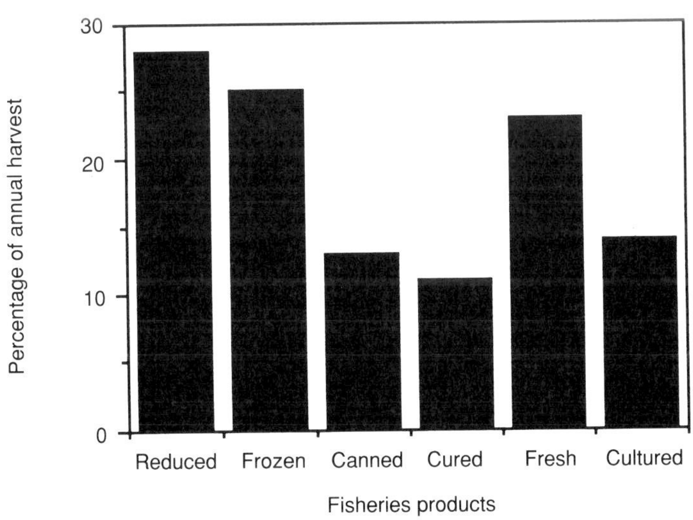

Figure 1.8 Cultured aquatic animal production (% of annual world harvest) compared with utilization (%) of all fisheries products (Source: FAO, 1994; 1992 data).

example, fisheries products as a whole also offer more alternatives for the consumers, and the markets of developed countries in particular are demanding variety and choice in their daily diets. Many individuals actively exercise and regulate their diets in the interests of their health, and at the present time most governments are promoting general campaigns to improve human health, and active campaigns by including plenty of fish in school meal programmes. Fish and fisheries products are highly nutritious, and any increased *per caput* consumption of fish and shellfish in any region benefits health. Aquatic animals contain a high level of protein (17–20%), with an amino-acid profile similar to that of meat of land animals. The flesh is therefore readily digestible and immediately utilizable by the human body. Compared with land animals aquatic animals (with some exceptions, such as shellfish), have a far higher percentage of edible flesh, and there is little wastage. Aquatic animals are also a source of minerals, such as calcium, iron, and phosphorus, as well as trace elements and vitamins. Marine species are particularly rich in iodine. The fatty-acid content is high in polyunsaturated fats, and particularly those which are said to reduce blood cholesterol.

Changes in consumer buying habits are also important. New and almost universal demands for 'fast foods', even in developing countries, provide opportunities for the institutional markets, together with pre-packaged refrigerated fish and fisheries products in new supermarkets.

Consequently, most governments give a high priority to aquaculture for food products for domestic markets in their development plans for food security and improved nutrition. This is equally important for land-locked countries with little or no access to fish, and countries with productive marine and inland fisheries which are perhaps already exploited at a high level. Countries which are not (and need not) give priority to aquaculture are generally those with natural fish resources which are as yet relatively unexploited.

In addition to the projected growth of national markets for aquaculture commodities, the high-value products (such as all crustaceans, marine fish, and some molluscs) continue to feature in the increasing level of fish trade. The prime international markets continue to be Europe, Japan, and North America, with marine shrimp, salmon, and prime marine fish (such as sea bass, sea bream, and turbot) in high demand. There is also a growing international trade in juveniles for supplying farms.

Another increasing international trade is in cultured tropical fish in preference to harvested wild fish. This can support government policies of conservation of natural resources and protection of certain environments. The trade is now a recognized form of earning foreign exchange. This is a practical objective for most governments for their aquaculture sectors, whether or not they have priorities for food security or improving nutrition. High-value export products, for example, do not, with some exceptions, provide a solution to domestic food security problems to any extent, and markets are more lucrative abroad.

Other social trends which may influence the growth of the aquaculture sector are internal population migration, tourism, and recreation. Again, in developed countries, there have been, in recent years, movements of populations towards warmer climates and particularly to coastal areas. These permanent movements, together with the temporary movements in the same directions through the growth of tourism, have meant increases in market demand for fisheries commodities in coastal resorts, and particularly higher value products which are new or different from those of the usual domestic diets.

Similarly there has been growth in the demand for recreational activities, such as fishing or simply keeping ornamental fish. Most countries provide organized outdoor recreational sectors and enhance natural waters with fish to create more tourist facilities, or to support national pastimes.

All these potential indicators and activities favour an increased growth of the aquaculture sector and its contributions to the growth of other individual economies. Favourable forecasts, however, are always subject to certain riders. For example, fish meal and fish oils are important ingredients in aquaculture feeds, as well as feeds for terrestrial animals and poultry. If capture fisheries are relatively stagnant, as noted above, then resources for fish meal are finite. Thus competition for this important resource will increase, with likely increasing costs. Fortunately, the constraint of availability of fish meal and fish oils, but possibly not the price, is not an immediate influence.

Increasing aquaculture production depends on the future of other resources, particularly fish meal

The production of fish meal world-wide is around 7 million t. The total amount of fish meal estimated to have been used in fish feeds in 1988 was 646 000 t, around 10% of all fish meal produced. The production of intensive and semi-intensive farmed fish is expected to increase by a factor of two to threefold by the end of the century. Demand for meal in their feeds is expected to double, with farmed fish taking 20–25% of world fish meal production by the year 2000. In other words, total supply of fish meal will be adequate to meet all needs of feeds for farmed fish.

Source: International Association of Fish Meal Manufacturers, 1989.

Similarly, the movement of people towards the coastal areas has caused land-values to increase and protected areas suitable for aquaculture to become scarce or too costly. Moreover, the residents and tourists like to consume high quality farm products but do not want the farms nearby, fearing the risk of pollution; and boaters complain of navigational hazards. Some governments find that the needs of aquaculture and the introduction of farm species conflict with their support for genetic conservation as a whole. In response, some governments, such as Hong Kong and Singapore, have gazetted specific areas for aquaculture only; while others, such as the State of Hawaii, have created coastal industrial parks for environmentally compatible 'green' industries.

In conclusion, the forecasting of the direction and strength of the aquaculture sector, like any other, depends greatly on the availability of statistical information. Good forecasting requires not only accurate information about the sector itself, such as the location and number of producers, types of species cultured, and systems and practices used, but also information about other competing sectors, particularly agriculture, as well as urban growth, tourism, recreation, and general social trends. Thus, assisting the process of planning at all levels is the primary function of collecting and reporting statistical data, and collecting and collating statistics is a vital government support service for managing the sector (see Chapter 7).

Chapter 2
Factors Which Influence Sectoral Growth

Synopsis

The chapter provides more background information about the aquaculture sector in general, in particular why and how its development is enhanced or constrained by factors not immediately concerned with the husbandry of cultured stocks. The first section deals with the natural environment in which aquaculture functions, and on which it is highly dependent. It notes particularly physical, chemical, and biological factors which may impact development, and describes their economic impact, either positively or negatively. The second section deals with the social environment, particularly the legal, economic, and cultural regimes. The third section briefly describes the constraints of the different levels of markets at which different products are targeted. Through illustrations, the chapter shows how these factors influence the selection of different farming systems, practices, and species for certain environments, as well as responding to the needs of communities and individuals.

In addition to internal factors which influence aquaculture development (as discussed in Chapter 1), there are many factors which are external to the technical processes of aquatic animal production, and may be either positive or negative in their impact. These factors also fit within a simple framework based on the broad setting in which the aquaculture sector is functioning. External factors are grouped in the framework according to their concerns with:

- the environment,
- the legal, economic, and cultural regimes of the society,
- the level of demand for the products.

The simple framework is illustrated in Figure 2.1, and the contents are discussed in the following pages.

INFLUENTIAL FACTORS		
Environmental factors	Social factors	Demand factors
(i) Physical • Competition for space • Competition for water • Human interventions • Industrial development (ii) Chemical • Water productivity • Toxic chemicals (iii) Biological • Population dynamics • Health and survival	(i) Legal • Ownership/rights to land • Ownership/rights to water (ii) Economic • Ownership/equity in operations • Ownership of livestock (iii) Cultural • Individual/community behaviour • Customs and traditions	(i) Local level (ii) Regional and national level (iii) International level

Figure 2.1 Factors which influence sector growth.

2.1 Environmental factors

Aquaculture is an industry which, like its close associate agriculture, is intimately integrated into the natural environment. It is recognized as a high risk industry predominantly because its many systems and practices often react immediately to changes which can occur among a great range of natural parameters, not only those of climate and meteorology. The magnitude or frequency of these environmental changes increase or decrease the risk of economic profitability of aquaculture accordingly.

Environmental factors which impact aquaculture can be grouped conveniently by their (1) physical, (2) chemical, or (3) biological character. This is because many environmental changes are due to the normal physical, chemical, and biological processes of nature, as well as the results of human intervention in the general milieu of everyday life. Many impacts of environmental change are positive for aquaculture, in that there is a direct economic benefit in some form; others are negative, in that risks to the industry are increased.

Any one impact caused by environmental change may not always be quantified accurately in financial terms, but all impacts influence the economics of aquaculture operations and management. Consideration of the consequences of environmental changes and their impacts is therefore an important part of the decision-making process in planning investment in the aquaculture sector.

2.1.1 Physical environmental factors

Physical factors to be considered for aquaculture

The two physical factors which are fundamental elements in all aquaculture production systems and practices are *space* and *water*. The two are best considered together as it is the close association of space with water which is important to aquaculture, and the physical contents of water (sediments and silts) rather than its mass and energy. Consequently, for aquaculture development, changes which occur to one or the other are directly linked.

The principal criterion of suitable space is its proximity to suitable water. There are some important secondary spatial criteria, such as the area available, gradient, geology, permeability, and accessibility, but these are very sitespecific characteristics.

For the many permutations of production area classes and potential production practices which are possible (see Figures 1.3 and 1.4, above), the space can be:

- inland or coastal areas adjacent to water,
- tidal flatlands or permanently submerged land,
- land above subterranean aquifers, or
- suitable rain-fed land.

Similarly the water may be:

- freshwater from any inland lake or reservoir, or any river and flood plain, or from underground, or good/low rainfall,
- brackishwater from any inland water body, or any coastal estuary or lagoon, or from any subterranean source,
- saltwater from any inland saltwater lake or coastal barrage, any nearshore or offshore sea, or any subterranean source.

Many of the high-value industries which compete for waterfront space, such as tourism, recreation, and select property development, demand large areas. The hotels and associated facilities (such as aquariums) of tourist resorts invariably occupy entire coastal bays, or extend along several miles of beaches, or entirely surround lakes. Typical recreational land-based facilities are municipal parks and golf courses (which may occupy 100–150 ha), and extensive national parks designed for natural conservation but open to the public. Many modern national parks are for coastal lands of natural beauty, and even underwater parks over entire coral reefs, all complete with visitors' centres and trails. In addition, typical water-based facilities are large reserved areas for boating, sailing, and water sports.

Other large tracts of land are now designated as nature reserves with restricted public access. These publicly owned areas are invariably to

Plate 1: Aquaculture facilities compete directly with development in other sectors for water and space. Plate 1a (top) illustrates a salmon hatchery in North America surrounded by increasing urban and recreational development: Plate 1b (bottom) shows an oyster processing plant now in the midst of industrial development in Hiroshima Bay (Japan).
Photos: Courtesy of Conrad V. W. Mahnken.

preserve ecosystems or wild and endangered species, and usually include habitats surrounding water.

In addition, continuous and significant demands on land and water are due to the normal consequences of population growth. Both are required by heavy industries and light manufacturing industries important to the economic development of any country, and these are rapidly increased by the subsequent demands for infrastructure to service them, and urban development for their manpower. The energy industry is particularly demanding of land and water. Nuclear and fossil fuel generating stations invariably have railheads and terminals for the delivery of fuels, and they also require large volumes of cooling water. Hydro-electric stations require the construction of barrages and barriers on rivers and lakes to store water, and these interventions are also an important part of the work of national water authorities to meet domestic demands for potable water as well as domestic water.

Other industries such as agriculture and fisheries compete for these same resources. Although much agricultural land is being lost to urbanization, new agricultural lands are being developed. This may be through the clearance of large tracts of forests for grazing land, as in Brazil; or through the transmigration policies of Indonesia; or through the drainage of wetlands in China and West Bengal in India. Agriculture development is often associated with irrigation projects, which require the construction of water barrages and reservoirs. The offshore capture fisheries industries require coastal land for ports and port infrastructure, including markets and processing plants; and coastal small-scale fisheries require small harbours and access points, as well as some shore-based facilities.

Aquaculture itself is also demanding of space and water, depending on which of the three production systems is selected (see section 1.4) and farming practice adopted. For example, the rapid growth in production of shrimp in the last decade has made extraordinary demands on coastal land and water use in tropical countries (see Figure 2.2).

Extensive production systems require little water and the most common farming practice is the simple individual earth pond relying on rain-fed water; however, these ponds often cover much space in one locale where farming is popular.

Extensive aquaculture systems are more demanding on space and less on water

In many African and Latin American countries earth ponds for subsistence or semi-subsistence extensive production may be as small as 1–3 ares in area and their production is small (about 200–1200 kg/ha equivalent). Although the majority of these ponds are random in size and shape, collectively their total areas are considerable as there are many farmers. In the Central African Republic, for example, there are reported to be 9000 farmers; in Côte d'Ivoire and Zambia about 2000–2500, each; in Nepal there are some 10 000. Many farmers own more than one pond.

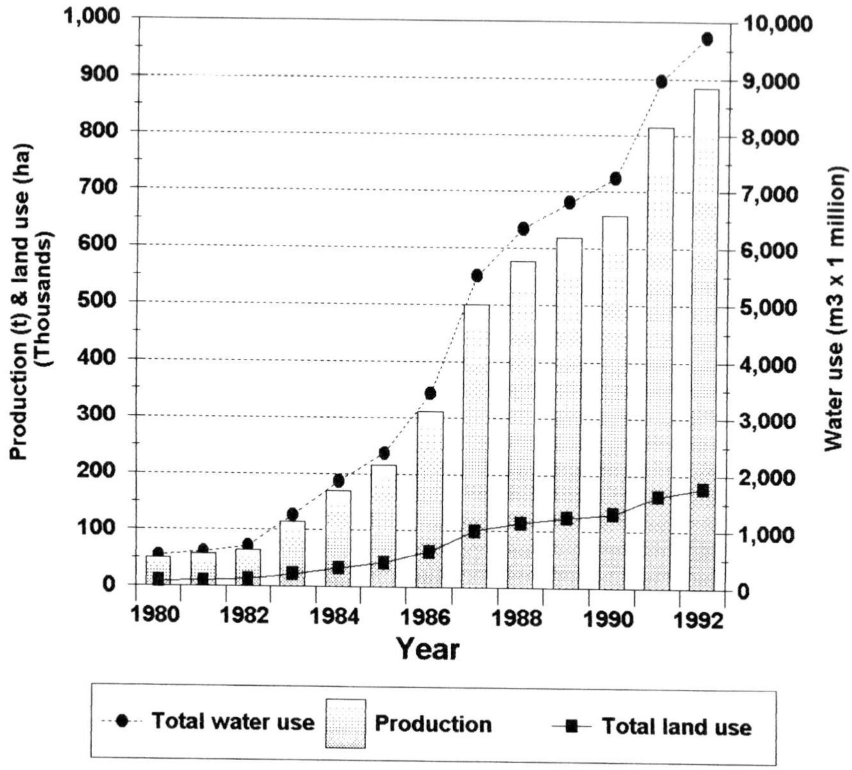

Figure 2.2 Increased land and water use through growth in global shrimp production (Based on average figures of 0.2 ha land and 11 000 m^3 water per t of shrimp produced).

Semi-intensive and intensive systems need constantly reliable resources of water, usually pumped from the ground into large ponds or funnelled through raceways. Although many farms may be located in the same area around a large water resource, in general both semi-intensive and intensive farms are less demanding on space as they are independent small-scale or vertically-integrated operations.

Semi-intensive and intensive systems are more demanding on water and less on space

Production in semi-intensive systems can be about 1500–8000 kg/ha equivalent, depending on water availability. Ponds, suitably constructed for the purpose, are typically between 0.5 ha and 50 ha in size, regular in shape, and together form farms which may be up to 500 ha in area. Typical water requirements for semi-intensive practices in ponds may be from one quarter to one complete change of water volume per day; intensive practices in elongated ponds, in concrete raceways, or in tanks, which may be quite small, may require from 2–10 changes a day, and sometimes even higher. Floating cages are one of the cheapest practices of an intensive system. Some cages are now constructed with volumes of 12 000 m^3, and surface areas of 2000 m^2, but typically they are from 50–200 m^3. Although this is small by comparison with earthen ponds, floating cages also require a considerable water catchment area around them. In Singapore, for example, a farmer may only occupy 30–40% of his leased area of 5000 m^2. Similarly net pens, which enclose areas in shallow coastal waters and lakes, collectively can cover large areas; for example, before management was introduced, over 25% of the 90 000 ha of Laguna de Bay in the Philippines was covered with pens.

In addition to these two physical factors of the environment over which certain industries compete with aquaculture, there are many physical processes of industries which directly affect aquaculture through the medium of water. These are the processes which change the quantities of sediments and silts in water.

Some increase in the level of sediments and silts may be due to natural processes brought about by climatic and meteorological changes; for example, in areas of volcanic activity, the seasonal land erosion characteristic of tropical countries with monsoon climates, the slow erosion and movement of sands typical of many coastlines, or floods and high winds. However, on the whole, the environment has adapted to these natural changes, and the aquaculture industry can operate around them.

Of more consequence to aquaculture are the unnatural increases in levels due to human interventions, usually through the same industries which compete for the resources of space and water. These interventions produce immediate impacts on aquaculture by changes in water quality from downstream siltation; for example, sand-mining in coastal areas and the extraction of gravel from river beds and estuaries which affect large areas many miles from the specific sites. Erosion following the degradation of agricultural land or through deforestation also produces an increase in sediments and silts which may impact on areas considerable distances away. In some cases the aquaculture industry itself contributes to erosion. Certain aquaculture sites, which use and discharge large volumes of water, can add to the sediment load and impact on smaller farms downstream.

Another physical factor due to human intervention is thermal loading. Certain industries use large volumes of water for cooling, such as the energy industry. The heated effluents create new micro-ecosystems within the vicinity of the generating plants to which many indigenous organisms cannot adapt.

Plate 2: Physical environmental damage to land and water resources seriously impacts on sectoral growth. Farmers, local village leaders, and extension agents discuss a comprehensive land-use plan for the plateau to prevent further siltation of the lower reaches of the Yellow River in China, which threatens fisheries and fish farming. The project is financed by UNDP and executed by FAO.
Photo: F. Botts, Courtesy of FAO.

Large volumes of cooling water can change the environment for aquaculture

The cooling process can add 6–12°C to the temperature of the water and, because of the volumes involved, large areas of the receiving waters are impacted environmentally and thermally stratified. A typical 1000 megawatt generating station requires some 2.5 million m^3 of cooling water per day, which is discharged back to receiving waters at the higher temperature.

The impacts of physical factors on aquaculture

All industries which compete politically, economically, and socially for the two primary resources of suitable land adjacent to suitable water have some negative impact on the development of aquaculture. Once any one of these industries is in place, the resources are invariably lost totally or severely constrained for aquaculture. There are some positive impacts on aquaculture through these same industries. A number of farms, for example, have benefited from the tourist industry by opening their facilities to visitors, or by building associated aquariums and visitors' centres. Others operate small 'put and take' fisheries. These are ponds designated for recreational fishing which are stocked with farm fish.

A major negative economic impact on aquaculture through industrial competition is the high cost of suitable land. Waterfront land is an important negotiable commodity because of its aesthetic appeal, and therefore its value does not conform to typical land values. Land available for development adjacent to the sea or large inland water bodies is invariably prime land sought by investors in high-value industries, such as select property development, tourism, and recreation.

Aquaculture sites compete for coastal land, the costs of which have soared

Suitable coastal land for aquaculture is often in high demand by other industries and land values have increased dramatically. For example, waterfront land values may be 3–5 times higher than cleared non-waterfront land, and 5–10 times higher than uncleared non-waterfront land, even though these non-waterfront areas may be suitably provided with subterranean water.

Land availability and cost are important factors for consideration before project investment. Aquaculture is predominantly a business for the individual entrepreneur. The individual cannot compete with corporate investors in the high-value industries, such as tourism and recreation, or with the socio-political motives of conservation. Consequently the high cost and alternative socio-political uses of waterfront land invariably prevents aquaculture investment altogether.

Aquaculture itself is also judged to have a negative effect on land values because of its lack of aesthetic appeal, and its industrial smells from processing and feed preparation. On the other hand, the aquaculture industry is attracted to remote areas where land and pristine conditions are still available. Consequently aquaculture has proved to be an industry which has assisted in the economic development and social stability of remote and rural regions, and prevented human migration into urban areas. Furthermore, not all farms manufacture feed or process the products on site, and are therefore a more appealing form of industrialization to integrate into these areas than many alternatives. Once in place, the aquaculture industry itself can protect an area and deter other less attractive competing industries from coming.

Changes to the quality of water through the presence of sediments and silts increase the risk to existing aquaculture operations. The loading of downstream waters with silts may directly affect the economics of farming operations. The impact may be no more than added maintenance costs, from such things as dredging accumulated material from ponds and channels, or unblocking filters, or replacing mechanical parts of pumps. Alternatively, the impact may be the increased risk to the health of the livestock, and potentially the loss of the stock. Sediment deposition in the ponds may make the substrate unsuitable for such things as nesting, and fine silts in suspension may impair breathing through blockage and damage to the gills.

Thermal changes have both a negative and positive impact on aquaculture. For aquatic animals and plants already at the upper end of their thermal tolerance, elevated temperatures are an added risk. Many of them are unable to withstand prolonged exposure to high thermal loading from industrial discharges, or the stratification they produce which may permanently change ecosystems. Animals and plants at the lower end, such as those of temperate climates, can tolerate exposure to higher temperatures better, and often have accelerated growth rates. This benefit can be used by aquaculture in controlled conditions, and there are many hatcheries and farms which use thermal discharge waters to advance the growth rates and reduce the time of production to market.

2.1.2 Chemical environmental factors

The chemical factors to be considered for aquaculture

The two principal chemical factors which effect aquaculture are *water productivity* as a result of nutrient loading, and the *presence of toxic chemicals* from pollution.

The productivity of water is measured by its biomass, that is the numbers of minute and microscopic plant organisms called phytoplankton. These organisms are the primary components in the food chain which, through photosynthesis and oxygen, link the chemical building blocks of life with life itself.

All waters which receive light have a natural productivity, but the two are not directly correlated. Many tropical waters, for example, in spite of their exposure to a high light intensity are not as productive as temperate waters. Variations in productivity are more directly related to levels of inorganic elements in the water, particularly carbon, nitrogen, and phosphorous, and certain trace elements.

Productivity levels are not constant. Continuous changes occur due to the physical disturbances of the water mass. Some of the most productive waters in the world are the regions where deep cold ocean currents rise to the surface over continental shelves bringing with them large quantities of nutrients. At the surface, with the increase in light intensity and temperature, the resulting biomass rapidly proliferates through the food chain producing rich fishing grounds. Such incidences of high productivity are not only confined to the oceans. Many shallow lakes in the tropics, for example, with nutrients kept in suspension by high winds and monsoon floods, have high productivities; and the waters of volcanic zones are invariably rich in nutrients.

Although productivity is influenced by natural changes, it is significantly influenced by human interventions. With the exponential rise in global populations and industrial growth, the productivities of both large and small water bodies are continuously changing through the increased utilization of water and the discharge of nutrients from industrial processes. Some industrial processes directly release soluble inorganic nutrients into the water, causing enrichment and eutrophication (an increase in primary production). Among these are wastewater treatment (which may release only primary or secondary treated effluents), agriculture (which produces wastes from animal husbandry and inorganic fertilizer residues in run-off), and aquaculture itself (which produces both animal wastes and residues of feeds). Other industrial processes provide conditions suitable for nutrient-biomass transfer. Among these are the energy industries which use large volumes of water which disturb a nutrient-laden substrate, and others which increase the temperature of receiving waters providing ideal conditions for growth.

There are natural environments where the levels of certain chemicals in the water are toxic to plants and animals, for example, where there are natural ores of base metals, such as lead, copper, and zinc. These are rare, and are of little consequence to aquaculture except in extreme conditions of drought (when they may be concentrated) or rains (when they may be dispersed into new areas). Acid sulphate soils on the other hand are common, particularly in large areas of coastal mangroves which are often reclaimed for aquaculture. As these areas are low lying there is high evaporation, which concentrates and increases acidity, as well as sodium and calcium salts.

The majority of toxic chemicals which can directly impact on aquaculture are the result of human interventions, and mostly transmitted through the medium of water. Industries which introduce toxic inorganic chemicals into water directly are the heavy metal mining industries. These include both precious and base metals, such as the gold and silver mines of Peru, the tin mines of Malaysia, and copper, iron, lead, and zinc mines of Africa and Latin America. Considerable volumes of water are used in wash-down, and the tailings are released into the natural water bodies or stored in quarries.

Other industries produce inorganic chemicals which create toxic conditions indirectly; for example, airborne sulphur from many industries using coal-burning fires is precipitated as 'acid rain', which concentrates in natural water bodies.

A number of inorganic chemicals are released directly into the water from man-made processes. Copper and lead, for example, leach from many protective paints, particularly marine paints, and aluminium from concrete. Aquaculture itself is an industry which uses both anti-foulants and protective paints, and many structures in concrete.

By far the most serious toxic chemicals which pollute water bodies and which directly impact aquaculture are the organic hydrocarbons. These bioactive compounds are invariably released by all their manufacturing industries, and again through their use, and their presence and effect are more lasting. Agriculture, for example, uses a variety of organic pesticides and herbicides which find their way through run-off into water. Algicides are introduced directly into water to keep reservoirs and channels free of algae.

Aquaculture itself also uses a range of chemical products which may be harmful to the environment. These include not only the more obvious algicides to keep the water systems free of algal growth, but also a range of chemicals used in operations to control diseases, increase production and survival, and produce a high quality product. Most of these are used as prophylactic treatments, such as organic fungicides and parasitides, or inorganic disinfectants and cleaning solutions.

The impacts of chemical factors on aquaculture

The majority of impacts of chemical factors are negative in that they directly affect the health and survival of the aquatic animals and plants being farmed through aquaculture. Any loss of production or change in the quality of the end product has a serious effect on the economic goals of the farm, and may result in bankruptcy.

In nature all aquaculture organisms are part of the dynamic food web. Productivity is therefore an important part of their environment and life-cycle, as they feed either directly on the phytoplankton and zooplankton populations immediately produced, or feed on them indirectly at higher levels in the food chain.

At times, however, productivity may be a risk. A productive environment, under certain conditions, can be become unsuitable and even lethal to the aquatic organisms living within it. The principal danger is an algal 'bloom' of particular phytoplankters, associated with diatoms, copepods, and ciliates. If nutrients are in the right supply, and conditions of temperature, salinity, and pressure are suitable, reproduction of these organisms becomes phenomenal.

There are four possible consequences of very high productivity, all of which are serious to aquatic animals. First, certain phytoplankters, particularly those belonging to the large group of blue-green algae (dinoflagellates), are directly toxic. If the population becomes so great that ingestion cannot be avoided, poisoning ensues. Second, the density may be so great that the breathing apparatus of the aquatic animals is physically blocked. Third, the oxygen demand of the population is so great that the environment is starved of oxygen; this usually happens at night before the process of photosynthesis begins again. Fourth, in a nutrient-poor environment following an extensive bloom, conditions are suitable for the reproduction of certain red algae, which cause the highly toxic 'red tides'. In nature, the majority of aquatic organisms can react independently to these unsuitable conditions of productivity, with the exception of sessile organisms, such as the molluscs (oysters and mussels) and algae, which are fixed to the substrate. The risk to them increases if the conditions are prolonged. In the farming conditions of aquaculture, there is little opportunity for escape, and toxic blooms and red tides have become one of the most serious risks to aquaculture worldwide.

Toxic algal blooms and red tides are probably the most serious risk to aquaculture

In Norway, in 1988, coastal algal blooms were responsible for the loss of over 25 000 t of Atlantic salmon, rainbow trout, in cage farms, together with some 5 million smolts. The estimated value of the loss was over Nkr 1100 million. Many farms were saved only by towing their cages away from the path of the bloom. Another 2000 t were lost in Chile.

Algal blooms also affect the quality of the finished product. There is an association between high concentrations of the blue-green genera with the 'off-flavours' of both fish and some crustaceans. Several incidences have occurred in the catfish farms of the USA when high numbers of large fish and associated high feeding rates have overloaded the ponds with nutrients producing outbreaks of phytoplankton. Deterioration in pond conditions resulted in an unacceptable product, and even mortality when there were associated temperature drops.

There are, on the other hand, a number of positive impacts through increased productivity. Many species of fish and shellfish feed directly on phytoplankton and zooplankton, and this is their sole food source. These are the species which are used in extensive farming systems, where the cost of inputs are kept to a minimum. Consequently certain forms of freshwater fish production are practised in places where productivity of the ponds can be increased by using nutrient-laden effluents.

Certain aquaculture practices are enhanced by wastewaters

In India there are large numbers of sewage-fed farms in many states; and wastewater-fed fish culture is practised in China, Germany, Hungary, Israel, and Peru. The use of treated wastewater has also been tested with marine species in brackish and salt waters, with good results. The growth of molluscs (oysters and mussels) is enhanced, and the products are depurated or cleansed before market. Because of the high levels of particulate matter in estuaries and coastal lagoons, the production of molluscs has been enhanced as well as used as a practical method to keep such areas clean.

The impacts of both inorganic and organic toxic chemicals on cultured animals and plants are much more direct and all negative. Most toxic chemicals affect the health and survival of aquatic animals and plants, and therefore jeopardize the profitability of the farm.

Toxic chemicals may be immediately lethal to cultured organisms, or poisonous after prolonged exposure; others interfere with many of the internal physiological processes, such as reproduction, metamorphosis, settlement, or moulting; others cause lesions and malignant growths which expose the animals to other diseases by lowering their resistance.

Other toxic chemicals affect the quality and hence the value of the product. They may make the product unsaleable if there are lesions and growths, or outward signs of ill health. Some affect the taste of the product. Although the human consumers avoid obviously affected and infected animals and plants, wild and domestic animals are not always protected. Consequently the effects of exposure to toxic chemicals are passed up the food chain, with the same consequences.

Some toxic chemicals make the environment unhabitable. Sulphur from atmospheric pollution and leaching from acid sulphate soils can increase the acidity of the water beyond tolerable levels. Sodium and calcium accumulations can change the salinity so that the ecosystem is totally altered. In the latter case, there may be positive impacts on aquaculture, as new environments are created which have aquaculture potential.

Suitable environments for aquaculture may be created by natural chemical changes

The inland salt lakes of Egypt are now sites for the culture of marine fishes and shrimps; salt pans are widely used for the production of live food organisms much in demand in shrimp hatcheries; and low-lying salt lands are unsuitable for agriculture and cheap to buy, and can be suitably dyked and flooded to reclaim for aquaculture purposes.

2.1.3 Biological environmental factors

Biological factors to be considered for aquaculture

The two principal biological factors which must be considered in the planning of aquaculture are the *dynamics of animal and plant populations*, and *individual health and survival.*

There are certain natural processes which interfere in the balance of animal and plant populations, but these are the long-term climatic and meteorological effects which are a part of evolution. It is the human interventions which are serious, some of which may be irreversible. Like many other industries, aquaculture has the potential to contribute to this interference.

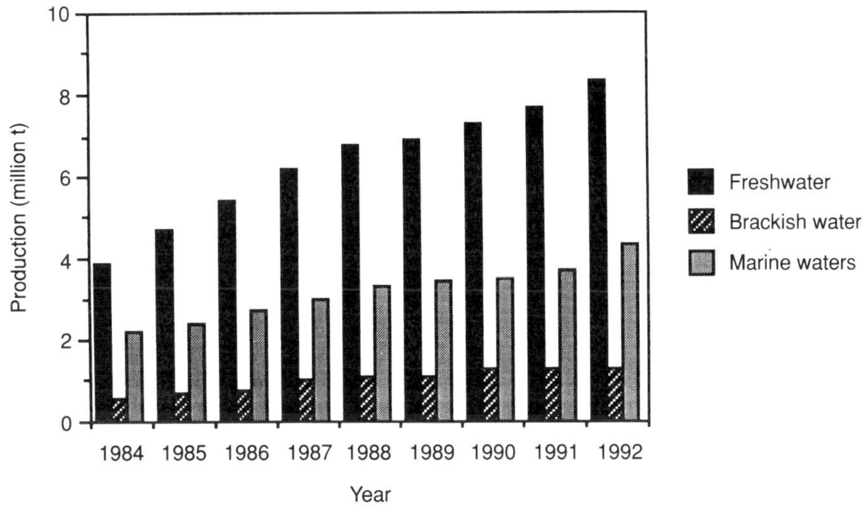

Figure 2.3 Aquaculture production (in million t) by aquatic environment (Source: FAO, 1994).

A principal requirement of aquaculture is suitable space adjacent to suitable water (see 2.1.1, above). The distribution of the present production of aquaculture in the three principal aquatic environments is illustrated in Figure 2.3.

In order to avoid the risks of pollution, many farmers are searching for suitable sites in remote and pristine environments, such as southern Chile,

---

Some aquaculture practices depend on the collection of wild resources

For stocking ponds, the farm marine shrimp industries of China (over 200 000 t) and Ecuador (over 100 000 t) depend largely on the collection of natural resources of post-larvae and mature female shrimp carrying fertilized eggs which are then spawned in hatcheries. About 70% of China's hatchery production of 60 billion post-larvae are from wild resources, but after 1990 a regulation required all production to come from captive stocks. In Egypt there is the annual collection of some 200 million juvenile mullet from the waterways of the Nile Delta. In the Philippines, before many coastal ponds for milkfish were converted to shrimp production, the estimated need for young milkfish was 1.75 billion pieces annually, although this number was probably never reached. In Taiwan the average annual collection of juvenile milkfish is about 90 million pieces. In Hong Kong there is the continuous collection of young groupers for stocking floating cages.

Plate 3: Many sectors currently impact environmental and conservation ideals. Plate 3a (top) the aquaculture research and development facilities of the Sea Fisheries Industries Authority (UK) located in the panoramic beauty of the Western Islands of Scotland; plate 3b (bottom) shows the development of luxury hotels adjacent to scallop farms in Shizogawa Bay (Japan). Lower photo: Courtesy of Conrad V. W. Mahnken.

or the far north of Norway and Scotland, or the smaller islands of Indonesia and the Philippines. Many of these remote places are natural habitats and often the last refuges of many wild animals and plants, some of which are protected. There are a number of potential consequences in the development of aquaculture in these remote areas. There may be a change or loss of the natural habit for wild animals and plants through the physical presence of an industry, however small, and its pollution. This may cause displacement of species, and increase the risk to some wild and endangered species.

Many aquaculture practices are still dependent on the collection of large numbers of natural fry. The loss of these resources has potential consequences for the dynamics of these and other populations of plants and animals in the ecosystem.

By far the most serious biological factor for consideration is health and survival. This factor must be assessed not only in terms of health and survival of the cultured species in question, but also for other species, and for humans.

Health and survival can be affected in many ways. First, there are some direct effects through disease. Large populations of animals and plants together, at times under stress, create opportunities for disease epidemics. Diseases of farm stocks may therefore be transferred to natural populations. Alternatively, certain cultured species of fish are hosts for human vectors, such as those of schistosomiasis and filariasis, and extensive areas of new ponds in tropical countries provide breeding grounds for malarial mosquitoes. It has also been theorized recently that integrated farming practices in Asia with pigs and fish are responsible for new variants of viral infections, such as influenza.

The transfer and introduction of new species to new environments provide opportunities for the transfer and introduction of pathogens. The movement of oysters and crayfish between USA and Europe in the past, for example, has been responsible for the introduction of diseases which have devastated the indigenous species.

Second, there are the consequences of genetic changes in both the cultured animals and plants, and the indigenous species. Cultured stocks, which may now be genetically selected or will be in the future, may escape and breed with indigenous populations. At the present time the attention of farmers is on genetic selection for characteristics which increase the value of the product, such as growth rate, size, and shape. Their selection procedures are often based on too few animals, and there is the potential for some severe genetic disorders if such individuals were to breed freely with wild species.

Finally, there are the consequences of consuming aquaculture products which are infected with parasites and pathogenic organisms, or contaminated with biological products used in farm and hatchery processes. Like animal husbandry, aquaculture has produced an array of veterinary medicines and biological products for many of its hatchery and farm processes. These include the more common antibiotics (such as penicillin, erythromycin, and streptomycin) to control bacterial infections; hormones (such as gonadotrophin, testosterone, and dopamine) to induce breeding; and stimulants (such as somatotropin) to enhance growth. There are also a number of natural products and artificial substances used to colour and flavour the flesh to ensure the final marketable quality of the product. Although obviously not harmful to the cultured organisms in the doses used, residues of these medicines and products can often be found in the environmental vicinity of hatcheries and farms. Through accumulation and concentration in the substrate, or in other organisms which are themselves consumed, these products may have far-reaching consequences in the ecosystem, particularly as a result of building up antibiotic resistance which might subsequently be transferred to bacteria of human and veterinary importance.

Residues of veterinary medicines and biological products may also be found in the aquaculture end-products which are to be marketed. Consequently an International Committee on Fish and Fishery Products (of Codex Alimentarius) is attempting to define a code for the quality and safety of aquaculture products, including safety and quality, water quality, off-flavours, residues of veterinary drugs, and the public health significance of diseases and parasites. Many countries already have their own regulations on the acceptability of food products containing antibiotics and hormones, and artificial colourings.

The impact of biological factors on aquaculture

There are many negative factors on aquaculture due to the practices of farming large numbers of individuals in one specific location. These include the result of individuals escaping into the natural ecosystem, which may have serious consequences on other economically important species. Many cultured species are carnivores and therefore predators on natural populations. Similarly, there is a negative impact on wild populations of important species which are lost or diminished through infection of farm diseases.

The practice of collecting large numbers of juveniles from the wild has a negative impact on aquaculture, although there is no economic evidence to support it. It is generally believed that the collection of populations of juvenile resources from the wild for stocking farms reduces the natural coastal and offshore fisheries. In view of the numbers of juveniles collected this is a reasonable assumption, but so far there has been no decline in the numbers available each year, or any decline in the offshore fisheries. The alternative (and objective) for aquaculture is the production of all its juvenile resources in hatcheries. At the present time this still requires considerable investment in research and development, and also in capital facilities.

On the other hand, some aquaculture practices contribute directly to commercial fisheries, and in some cases are the basis of these fisheries. The practices of stocking or restocking (the release of juveniles into closed water bodies) and enhancement (the release of juveniles into open water bodies) with valuable commercial species contributes significantly to many inland fisheries, and several coastal fisheries. These practices also support other fisheries, as they provide food for other species, and to wildlife in general, including mammals and birds. Some endangered and protected species, such as sea otters and sea eagles, have benefited from the presence of fish farms within their territories. Furthermore, alternative resources of cultured commercial species may also reduce the pressure on many wild populations, currently overfished, which allow them to recover to support a more managed commercial fishery.

There are many negative impacts on aquaculture due to the introduction and transfer of species. Like agriculture many decades ago, the introduction and transfer of non-indigenous species is playing an important role in the development of aquaculture worldwide. The negative impacts are not easily assessed in economic terms. There have been losses or reductions to commercial species through the associated introduction and transfer of disease. There may also have been the loss of certain genetic resources in native populations following interbreeding with farmed stocks. On the other hand, many countries now have thriving commercial sectors based entirely on non-native species.

New aquaculture industries can be created through introductions

The global freshwater prawn industry is now over 20 000 t annually. Much of it started from an introduction of 15 adults from Thailand to Hawaii in 1962 for breeding and culture; a mass-propagation technique was successfully developed, and juveniles transferred from Hawaii all over the world. The annual production of Pacific and Atlantic salmon in Chile is now over 30 000 t; this follows from a number of introductions made since 1974.

The risks to human health have potentially many negative impacts on aquaculture. Fortunately, the majority of these risks can be readily avoided, otherwise the aquaculture industry as a whole would not be economic if its products were unsaleable, or its environment dangerous.

In many cases, the economic impact on aquaculture is the cost of diligence. For the parasitic diseases of the tropical farm environment, such as schistosomiasis, filariasis, and malaria, etc., can be monitored by management and controlled by biological and chemical methods. In addition to health monitoring of farm staff, there would also be the health monitoring and possible treatment of communities within the district. For viral and bacterial diseases through consumption of infected products, such as infectious hepatitis, typhoid fever, gastroenteritis, and cholera, etc., the need is for correct handling, processing, and storage of all commodities, and final preparation before consumption.

The presence of residues of veterinary drugs and biological products can be eliminated or reduced by appropriate use. However, monitoring should also be carried out by management. The presence of health hazards in particulate matter, especially in shellfish, can be eliminated by depuration and cleansing. This is an additional cost of post-harvest processing. In fish cultured intentionally or exposed accidentally to sewage or wastewater reared it can be avoided by keeping the fish for several days in water of high quality, followed by careful handling and processing (washing, cleaning, and storage at the right temperatures). The crustaceans are the most difficult to clean, and therefore considerable attention must be given to processing, and storage. The consumer is further safeguarded from all these potential risks by the public health services which set standards and codes of practice for product quality and handling, and which continuously monitor food products.

The risks to public health which are the most difficult to detect and monitor are the phycotoxins produced by the normal physiology of the cultured animal. In shellfish there are several toxins responsible for paralytic, diarrhoeic, and neurotoxic shellfish poisoning; in certain tropical fish there are ciguatera toxins. The toxins are produced by the aquatic animals after consuming certain species of phytoplankton, particularly armoured dinoflagellates, but not always during blooms or tides. All these toxins are dangerous to humans, and occasionally lethal. There is no way at present to detect fish which have ciguatera. The incidences of paralytic shellfish poisoning are more likely to occur during the seasons of blooms and tides, and the necessary precautions of depuration should be followed.

In spite of these small but nonetheless potentially negative impacts on aquaculture through consumption of its products, there are some positive impacts. The protein components of fish flesh, for example, are readily digestible, and fish have a far higher percentage of edible flesh compared with terrestrial animals. Aquatic animals are also a rich source of minerals, trace elements, and vitamins. The benefits of aquatic animals on human health have been noted in section 1.5.

2.2 Social factors

In addition to factors of the natural environment which may influence the development of the aquaculture industry, there are a number of factors which are best described collectively as 'social'. These include the legal, economic, and cultural regimes of the society in which aquaculture development is required to function. These regimes are best considered together as they are closely linked, and collectively encourage or compel national producers to adopt one type of management approach to aquatic animal production over another.

Social factors influence different management approaches at a national level

In Israel the collective settlement policies and social practices have resulted in the development of large farms owned and operated by the kibbutzim to produce food-fish to meet community and national needs. Communal township farming in China is the backbone of production of large quantities of similar freshwater fish for local consumption, whereas in Taiwan the national framework encourages free enterprise among individuals resulting in large numbers of small-scale family-owned farms producing a range of high-value products, many for export markets to earn foreign exchange. Countries with centrally-planned economies only operated, until recently, through cooperatives and state farms.

The legal regime of each social system, on the whole, has the greatest influence on the development of the different management approaches (described in section 1.3) to meet national and individual goals. This is because in some countries national legislation is structured on common law (through precedent) and in others it is on civil law (through decree). Furthermore, some countries have a federal structure which entails a distribution of authority between the central government and the state or provincial authorities. Economic and cultural regimes influence more the selection and adaptation of different systems and technical practices to meet national and individual 'returns on investment'.

2.2.1 The legal regime

At all levels of the aquaculture sector the legal regime influences decision-making, whether by public administrators or individual producers. It should be noted that the use of legislation as a government policy instrument to regulate and control the aquaculture sector is the subject of Chapter 6, and therefore legislation governing farming activities is not considered in detail here, but only the broad aspects with regard to sector development are summarized.

It is important to recognize that the nature of any legal regime might have little specific relevance to the aquaculture sector *per se*. Some countries, for example, with an advanced aquaculture sector, have virtually no direct legislation on aquaculture, but regulations are to be found in laws regarding land, water, and environmental control. At another extreme, it is possible that a country has an elaborate legal regime for aquaculture, but very little aquaculture activity so far. On the whole, most developing countries have little legislation specific to aquaculture and many developed countries are probably already over-regulated.

Differences within any one country may also exist, depending on the environment in which the activity is carried out, or the production systems and practices selected.

Aquaculture practices are subject to different areas of law from country to country

Many marine production activities, such as sea cage farming and sea ranching, may well depend primarily on the general marine fisheries law (which may or may not address such activities specifically). Inland aquatic farming in freshwater, on the other hand, which utilizes public lands, may depend primarily on laws affecting the public domain or water resources, and on private lands it may depend primarily or even exclusively on land laws, water laws, and environmental laws.

One important aspect of the legal regime for aquaculture development is that it is almost entirely national in scope. Unlike the legal regime of marine capture fisheries, international law issues have only a marginal impact on aquaculture development. The international legal regime must be considered for aquaculture activities in inland lakes, rivers, and estuaries (or possibly a water table) which border countries; in coastal areas, including their extended economic zones (EEZs), where navigational rights are important; and for ocean ranching of anadromous or catadromous species. Thus the extent to which a legal regime already exists, or how it may be in the future, is specific to every country and part of its domestic law.

In terms of general sector development the legal regime of concern to policy-makers and planners (and of course investors) is the access to and use of resources. There are three specific issues of importance to aquaculture; these are:

- ownership or rights to the land, both onshore and offshore,
- ownership or rights to water, and
- concern for the environment.

It is obvious that aquaculture is therefore subject to many ancillary changes to the laws regarding the resources of land, water, and the environment, which may be made quite independently and for other reasons.

Ownerships or rights to land

Many countries at present pay little attention to the use and access of land devoted only to aquaculture. However, development may be constrained by laws regarding public and private property rights, agricultural development policies and land reform acts, environment protection policies, and national security and defence considerations (see section 6.3).

As many aquaculture operations are carried out on land in the public domain, such as coastal areas, or riparian lands, they are subject to special concessions or authorizations granted by authorities with vested responsibilities. Concessions, through licences or lease agreements, are usually granted for a period of time only, and cannot be transferred or sold, and may be limited in size. In order to increase investment in the sector, many countries, such as Colombia, Panama, and Venezuela, have policies and laws which encourage access and use of any public land; in other cases, such as Egypt, it is not encouraged except where the public land is unsuitable for agriculture.

Ownership or rights to water

Water is obviously the most essential resource for aquaculture, together with almost all other industries, and therefore most governments, at all levels, have legislation regarding its use and misuse (see section 6.3.2). Differences between countries occur regarding the legal status of ownership of water, the explicit reference to the need of a licence for aquaculture, and the place of aquaculture in the order of priority established among beneficial users. For example, in some countries, such as Zambia, ownership of all water is vested in the state, and its uses are classified. In Panama the primary concern of the law is the conservation of all water and its optimum use, followed only by other uses.

In most cases national legislation and priorities are in regard to freshwater only, but in some cases it extends to brackishwater and even marine waters. Unfortunately the responsibilities regarding access to and the use of water are vested in several government ministries or ministerial departments. The regulations with regard to right to use water for aquaculture are invariably part of the use of water for fishing or agricultural purposes in general, but now some legislation is specific to aquaculture. Similarly regulations concerning the discharge of waste water and effluents from aquaculture sites, sometimes with very specific quality aspects, can be found. Thus most countries regulate the amount of water available for aquaculture (and other) purposes through permits, licences, and other authorizations.

Concern for the environment

Most countries have some level of legal concern for the environment (see section 6.4). As noted in section 2.1 aquaculture is highly integrated with and impacted on by the environment but, in turn, a number of aquaculture activities themselves can have adverse impacts on the environment. Thus aquaculture is influenced by legislation governing pollution by domestic and industrial wastes, and by pesticides. It may also be governed by legislation dealing with human health and welfare (as it produces food for human consumption, and farm conditions may encourage the transmission of vectors), and by legislation which protects certain ecosystems (such as coastal tidelands, mangrove swamps, and estuaries). Of particular interest to aquaculture are environmental aspects dealing with:

- land use planning and zoning,
- control over installation and operation,
- discharge limits and pre-discharge treatment to meet limits,
- fiscal incentives, and
- prohibitions.

Most countries also have legislation regarding the spread of fish diseases and the movement of live fish, particularly the introduction and transfer of marine and freshwater organisms. Considerable leadership in this direction has been given by the European Inland Fisheries Advisory Commission, which in 1988 published Codes of Practice and Protocols for its member countries. These have also been taken up by other regional fisheries commissions, which, although not legally binding, are the accepted basis in most countries.

Other international bodies are also providing guidelines within their technical purview. The multi-organizational Joint Group of Experts on the Scientific Aspects of Marine Pollution (GESAMP) has proposed a number of activities for reducing environmental impacts of coastal aquaculture.

International organizations provide guidelines on reducing environmental impacts of aquaculture

'Specific actions are essential to effectively utilize the environmental capacity of the coastal ecosystem for food production and generation of income, reducing resource use conflicts, and minimizing health risks to human consumers and adverse ecological impacts. These activities include the formulation of coastal aquaculture development and managerial plans, application of environmental assessment to aquaculture proposals, development of criteria for site selection, determination of the carrying capacity of ecosystems, establishment of guidelines governing the use of mangrove wetland, bioactive compounds, transfers and introductions, improvement in farm operation and management, regulation of farm discharges, and monitoring ecological changes, and application of regulatory measures and economic incentives or deterrents to promote sound environmental management'.

Source: GESAMP, 1991.

2.2.2 The economic regime

Development of aquaculture is influenced by the economic regime of the country, which may be either a market economy or a centrally planned economy. Neither type of economy provides a constraint to aquaculture activities as a whole, but examples of sector development under both regimes indicate a wide disparity in production and efficiency, depending on the systems and practices employed.

National economic frameworks influence the basic organization of the aquaculture sector

For the most part free-market economies are characterized by large, corporate-owned, vertically integrated farms, and small-scale owner-operated farms, such as are found in Ecuador, Japan, Norway, the Philippines, Taiwan, Thailand, and the USA, etc. Centrally planned economies are characterized by cooperatives and state farms, such as are found in China and Cuba, and in old Eastern bloc countries of Europe, such as Albania, Czechoslovakia, Hungary, and Yugoslavia.

In terms of general sector development the economic regime of concern to policy makers and planners (and of course investors) is ownership. There are two specific issues of importance, namely:

- ownership or equity in the farming operations, and
- ownership of the livestock.

Ownership or equity in the operations

All private investors, whether individuals, equity partners (including private cooperatives), or publicly owned companies, require total control over their investment, and therefore ownership of the operations is understood. The farm products are invariably high-value species for luxury markets, and even exports.

Public investors, on the other hand, either through state cooperatives or state farms, consider that ownership remains in the hands of the state. The farm products are usually middle-value and low-value food fish geared to national food security, or for stocking other inland waters. Most state operations are not independent financial cost centres but are subsidized by government as there are often price controls on the products. Recently, the state cooperatives and farms have become more independent, particularly in the countries of Eastern Europe and the new independent states of the former Soviet Union, and managers and the work-force share in any increased profits through bonuses, which may be in kind. Many community townships, particularly those in China, are now being broken down and ponds distributed among the members to increase production. In these cases the products may still be marketed through the township to maintain control, but in others they are not.

Ownership of the livestock

Both private and public investors, in principle, require total ownership of the livestock on the farm. However, at the present time there are some anomalies in certain types of practices. For example, many farms are still dependent on the collection of wild seed from natural resources in the public domain, such as juvenile shrimp for farms in Ecuador, milkfish fingerlings for farms in Indonesia, the Philippines, and Taiwan, and mullet fingerlings in Egypt and Israel. Their collection is often the sole economy of coastal communities and, through middlemen, sold to the farmers, thus becoming their property. Some farmers, such as the oyster and mussel farmers of France and Spain, produce their own seed but grow them out in the public domain on floating rafts or in managed tidal beds. However, they still maintain ownership.

The most difficult issue regarding livestock ownership occurs with farmers who ranch migratory anadromous and catadromous species, such as salmon and eels. They produce fingerlings in private hatcheries but again grow the fish out in the public domain. Many are caught by fishermen, as they are indistinguishable from wild fish, and they only become owned by the farmers when they return to the point source of release. However, in time, genetic markers and other means of identification will be developed to prove ownership at all times, and therefore there are likely to be problems with the fishermen who take them at sea, together with the other fish which are cultured and released to enhance the natural fisheries.

2.2.3 The cultural regime

The third important social factor which influences development of the aquaculture sector is the cultural regime, specifically the customs and behaviour of individuals and communities with each other, and their collective attitudes towards farming fish and consuming fish.

Certain regions of the world, such as North and South America, and the Eastern European countries, are not in general, fish consumers. The people prefer red and white meats, and the countries have highly developed farming industries supplying these commodities. Consequently their fisheries and aquaculture sectors are small, and in some cases non-existent, and they find it difficult to compete and change traditional eating habits. Most peoples of Asia, on the other hand, are high fish consumers, and some countries have extensive fishing fleets and large aquaculture sectors to meet the demand. The aquaculture sector in many countries is advanced, as farming fish has been traditional for centuries. Western European and Scandinavian countries are also high fish consumers but have access to rich fishing grounds. Consequently their aquaculture sectors are active but still relatively small and specialized. Island peoples, in general, are also fish consumers, but are supplied by their natural coastal resources. Thus aquaculture development is most rapid where there is a tradition for fishing and consuming fish, although it may be restricted in scope. Only Asian countries have the potential for producing a very broad range of species through a variety of systems and practices; most other countries are initially confined to a small group of middle and high-value species which can compete with other products from fishing, and with red and white meats, until the sector becomes established.

Individual and community behaviour

Closely allied with the consumption of fisheries products to advance development of the sector are the customs and behaviour of individuals and communities, particularly with respect towards both fishing and farming, but also with respect to work responsibilities within the household and community structure.

There is some evidence that individuals and communities who fish have an affinity towards farming aquatic animals and plants. In Japan, for example, many fishermen have successfully practised aquaculture as a second job. However, in general, good fishermen do not necessarily make good farmers. For the most part, the aquaculture sector is dependent for growth on a new breed of entrepreneurial individuals, and in sufficient numbers to provide a critical and sustainable mass.

Initially the most logical approach has been to encourage terrestrial farmers to take up aquaculture, but obviously this is restrictive as there are many options for aquatic farming. Rural subsistence farmers have been encouraged to include fish farming within their range of daily activities to provide family food sufficiency, but this has not worked well. Fish farming requires considerable more attention as the farm requires constant supervision and this the farmer and his family have not been able to give. Rural societies where it has worked reasonably well are those where women have actively taken responsibility and may own (in exchange for work) all or some of the farm ponds, and/or own and sell the products, such as in parts of Asia (China and the Philippines) and to a small extent in Africa (Côte d'Ivoire and Zaire). However, this is not the case if men still own the ponds and/or the products, or control the selling.

Many rural subsistence development projects, including those for displaced fishermen, have generally found that a few individuals rapidly dominate the scheme, buying out the ponds or cages of other farmers. This is indicative of the entrepreneurial nature of aquaculture, and illustrates that successful development is very dependent on a core of these types of individuals and their families. Integrated farming, raising fish in association with animals and poultry and even crops, is one approach in particular which is highly effective, as the range of products adds substantially to the household income and allows for greater part-ownership within the household.

Individual farms and integrated farms have also been beneficial to communities. The most progressive, as noted above, are the kibbutzim of Israel where the farms are one component of an organized economic unit, even to the extent of undertaking applied research. Less sophisticated are the communal townships of China, which tend to base their economic unity around one type of product, such as a shrimp farming commune, or products which are readily produced through close integration, such as fish and ducks.

Individuals and communities in countries which are generally market oriented and entrepreneurial in outlook, such as the Philippines, Taiwan, and Thailand, have few constraints for developing a sector, other than capital. If there is potential for a high-value commodity, such as shrimp in tropical countries or salmon in temperate countries, then frequently foreign capital can be readily found (for example, joint venture investments in Ecuador and the Philippines for shrimp, and in Chile for salmon). The availability of investment and consumer capital is important. Free market countries in Africa, on the other hand, have been constrained by the lack of capital, and where aquaculture has developed production has had to focus on low-value products for local markets and the low incomes of the population. Thus, compared with aquaculture development in Asia, development in Africa, and to some extent in Central and South America, has been a greater risk for farmers.

Customs and cultural traditions

Depending on the type of farming activity certain customs or cultural traditions may present a barrier to development. Issues for subsistence farming, particularly in African countries, invariably concern the traditional place of women in the household and their rights or lack of rights to own land or ponds, and own and sell the products. In some countries the men farm and own all animals (and fish), and women the plant crops. In certain Polynesian societies of the Pacific all fish and fish ponds belong to the chief, and the men of his village do the work. All women are excluded and even forbidden from working in fish ponds, and even fishing other than collecting shellfish off the beach. In countries of Asia and the Americas there are few such barriers, and women actively participate in farming and own farms independently.

In conclusion, the existence and extent of the legal, economic, and cultural regimes in any country, state, or local area, and any changes which may occur, collectively influence the development and direction of the sector. In particular they influence the type of aquaculture which can be developed, and the systems and practices which merit investment. In many respects, their influences are already becoming well established but at greatly different levels throughout the major regions, as illustrated by the current geographical disposition of aquaculture production world-wide (see Figure 2.4). Compared with the countries of Asia, Greater Europe, and North America, for example, the social regimes of most countries of Africa, the Middle East, and Central and South America, are still not conducive to rapid sector growth.

As described at length in Chapter 1 (section 1.3), there are three primary groups of aquaculture activities which are options for the public or private investors to pursue. These are:

- aquatic farming, i.e. total management of all processes, including propagation, grow-out, harvesting, and marketing of products for human consumption;
- enhancement of natural fisheries; and
- culture of non-food products.

Pursuit of any of the different activities within each group must be acceptable within the legal, economic, and cultural regimes of communities and individuals potentially making up the sector as producers and consumers. The influence of these regimes, either individually or in combination, on these activities, and their effects on the respective types of producers, are the reasons for the diversity of aquaculture. Thus it is important that planners and administrators of aquaculture sectors select, promote, and develop the systems and practices which are applicable to the social framework of their country as a whole, or to that of specific regional and local areas.

2.3 Demand factors

Aquaculture, as noted in section 1.2, must be profitable to the investor. However, profitability depends on the existence of a consistent demand for the products, and a thriving aquaculture sector cannot be established until this demand is being met regularly and reliably by producers with acceptable products at the right price. The products, in one form or another and fairly priced, must therefore be appealing and popular among a significant number of consumers, either locally, nationally, or internationally. Consequently, identifying and meeting the needs of these consumers, within the economics of production and post-harvest operations, are important influences on sector growth. They are the responsibility of everyone in the industry.

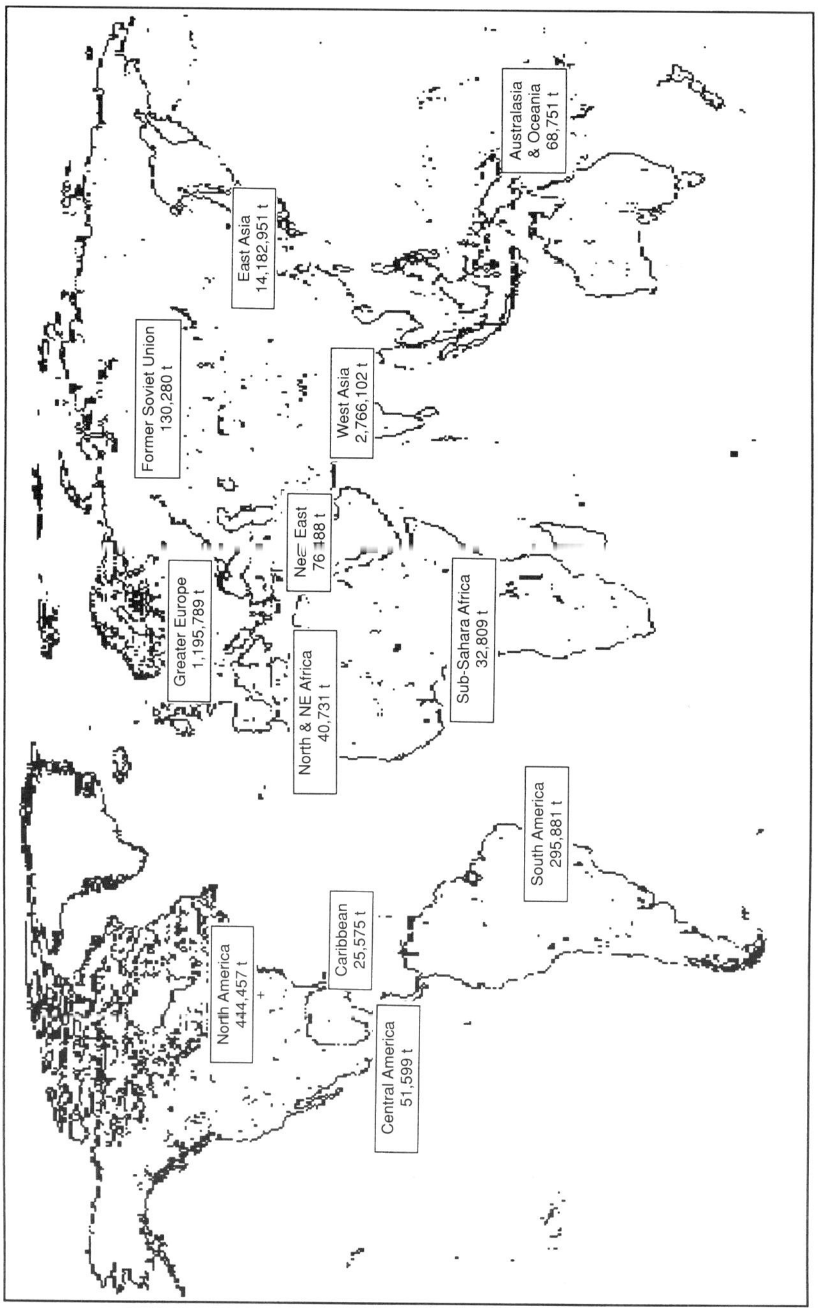

Figure 2.4 Regional distribution of aquaculture production (in million t, 1992 data; Source: FAO, 1994).

Demand at the local level

At the local level, demand is little more than seasonal preferences and customs characteristic of the local population who use the weekly markets of the nearest village or urban centre, and frequently interspersed with religious festivals. This demand is met by small-scale and subsistence farmers, typical of remote rural areas of Africa and Asia, who have few resources and selectively harvest fish for market (or barter) each week.

Consumption and marketing structures are often very traditional in these remote inland areas, and farmers simply adapt to the local system and provide fish (such as tilapias, carps, and other freshwater fish cheaply produced) to keep within the small purchasing power of the local people. Preferences are for live, salted, or dried fish, although large fish are often purchased in small pieces. Typically the women in the farmers' (or fishermen's) extended family will be responsible for marketing, and will transport and sell the products, taking only what they can carry to the local markets; alternatively, the products may be sold directly to consumers (other villagers), or bartered for agricultural produce and other needs. Farm-gate sales are also carried out directly with buyers, or 'fish mammies', who then act as vendors or brokers between market and street vendors. In some countries, such as Nepal and northern India, fishing and selling fish have been the traditional cultural prerogative of certain tribes or social sects, and these same people are now contracted by the fish farmers to catch and sell their products.

Estimating the demand at these local markets is mostly by trial and error on the part of the producers and traders, as no factual information is available. The total local demand is not significantly large, and therefore the risk to the individual farmer is reduced by the fact that not all his crop is being harvested at any time.

The protein needs of rural populations are frequently only provided by harvests of natural fisheries, supplemented by aquaculture products. However, the growth of an aquaculture sector in the country will always be constrained if development is restricted to the local level, primarily because it has little economic importance and therefore priority for organization and management (see Chapter 5).

Demand at the regional and national levels

Demand at the regional and national level becomes a much greater risk for the individual owner-operator. This higher industrial level, with the quantities involved, can only be met by farmers who have large resources of fish and shellfish of similar size and age dispersed throughout a number of facilities. Here, the entire crop of a pond or cage must be harvested at the same time, and immediately processed and sold.

At this level, consumers are concentrated in the larger urban centres or into institutional markets, and therefore marketing large volumes of a product is potentially very feasible. They also have more purchasing power and are more demanding on choice, and consequently farmers can produce higher value products, choosing from a wide variety of species within their technical skills and the local environmental resources. These options are frequently marine and brackishwater resources (such as shrimps and prawns, milkfish, mullets, sea breams, sea bass, groupers, etc.), but also include many freshwater species (such as trout, Chinese carps, eels, and crayfish, etc.).

Fortunately, the potential for marketing this greater variety of products in large complete crops does not necessarily rely totally on individual producers. Between the producers and the consumers at this industrial level are critical operations of harvesting and post-harvest handling. These require a range of marketing services, managed by others who act on information frequently far removed from that which influences producers.

There are several options for the producers at the industrial level to interface with the necessary operations of harvest and post-harvest. These include:

- a producers' cooperative,
- an organization of producers,
- the owner/operator of the nucleus estate,
- establishing a vertically integrated company,
- selling directly to a processor or broker.

Meeting consumer demand at a regional and national level therefore relies on producers and a large group of secondary-level players providing a variety of supporting services. Depending on the specific farm products. these may include live-fish haulers, ice-manufacturers, processors, packagers, storers, distributors, and wholesalers. Creating and sustaining this secondary support level has historically been one of the major constraints to sector development in many countries, as it is dependent on a large volume of high-quality products. These must be supplied regularly and reliably from a few very large farms, or from large numbers of small farms, over a prolonged period of time, year after year.

While a critical mass of producers is essential for the basis of an aquaculture industry, the involvement of these supporting services, particularly processors, is vital for the sustainability of the sector as a whole. Processors are essentially the primary market, or the immediate interface, for producers attempting to meet regional and national consumer demands. Moreover, they benefit from further consolidation and growth of the industry by introducing quality control standards for both producers and consumers. These standards improve product quality for the consumer, and stabilize prices paid to the producer.

Consumers demand products reliable for their quality, and at a fair price

Processors have contributed to the sustainability and growth of the aquaculture industries in many countries by imposing standards and quality control. For example:

- In Norway, and countries of western Europe, Atlantic salmon and rainbow trout are graded for size at harvest to prepare them for processing for specific markets.
- In Indonesia, and other countries of Asia, marine shrimps and prawns are immediately chilled and iced-down for transportation and processing to preserve texture and prevent bacterial contamination.
- In the USA, catfish, and some other bottom-living fish, are first tested for 'off-flavours' before being accepted for processing.

Demand at the international level

International markets are theoretically the consumers with the greatest purchasing power. Potentially they provide the greatest return on investment for the producer. This may be directly, through high prices, or indirectly, through investment incentives (see section 7.3).

For the most part, the international consumer demands for fisheries products, particularly high-value sea-foods, are well identified and have been supplied historically by natural resources sold through well-established marketing structures. At first, aquaculture products (such as marine shrimps, Atlantic salmon, and rainbow trout) have simply provided an alternative resource for these structures. However, in time, farm products proved consistently more reliable in quantity, size, and quality, and have enabled certain markets to expand, or have captured a larger share of the markets displacing the natural resources.

For both producers and post-harvest service suppliers, developing an aquaculture industry to meet consumer demands of international markets is mainly an extension of the risks encountered at the regional and national level (above), but with additions. Most new risks are concerned with acceptability of the products, particularly if the importing country has its own domestic market. These may include more rigid quality control parameters to conform with certain food, hygiene, and health standards imposed by the buyers and the regulatory authorities in the importing country; for example, artificial colouring of the flesh, residual veterinary medicines or drugs, and the levels of different micro-organisms. Other risks are associated with the terms of the contracts, particularly regarding delivery, price, and receipt of payment.

International markets for the same products differ from country to country, and particularly when a domestic market already exists

Japan is a selective importer of farmed salmon, mostly for a growing restaurant trade. Coho salmon, from Chile, were as high as US$6.2/kg in late 1993, but dropped to US$4 by January, 1994. The traditional retail outlets in Japan prefer the wild resources of Pacific salmon, particularly sockeye, king, and coho salmon imported from Canada and USA (Alaska), and their own coastal resources of chum salmon. In late 1993, frozen Alaska red (sockeye) salmon from Bristol Bay were wholesaling for US$6.5/kg for large fish. Retailers also prefer troll-caught fish. As with all general trends in fish consumption, different market peaks also exist from country to country with regard to seasons, national festivals, and religious observances.

Because of the risks involved, the international markets for aquaculture products have only been exploited by large vertically integrated producers, nucleus estates, and small-scale producers working through efficient producer organizations. These have been supported strongly by the respective government department(s) responsible for the development of exports to earn foreign exchange.

In conclusion, the demand for fish and shellfish by consumers is already well-established in most areas, having been met by fishermen harvesting natural marine and inland fisheries resources. Although this traditional demand has simplified the marketing of aquaculture products, the competition between fishermen and aquatic farmers of all types (see section 1.3.1), and amongst an increasing number of farmers, is exacerbated by the fact that most of them offer consumers the same products. An understanding of consumer preferences and habits, and attention to quality and service, by individual producers are therefore of paramount importance for an emerging aquaculture industry to be sustained and steadily expanded.

Chapter 3
The Sector Planning Process

Synopsis

The chapter describes in detail the many components involved in sector planning. After a brief introduction on the importance and need for sector planning, the first section describes both broad and specific problems which often arise, and key aspects of planning, particularly choosing the right framework and the time horizons of plans. It makes specific reference to the classification of plans. The following section deals with planning at the regional level, and its problems, and emphasizes methodology and organization.

3.1 An introduction to sector planning

3.1.1 General background

The planning of aquaculture development is not yet widely practised. Even in developed economies it is relatively new, and not comprehensive. The great diversity of the sector (illustrated in the previous chapters), and its rapid modern development in the last thirty years, have necessitated the development of national government policies. In most cases these have been instituted to promote further growth for the sector while at the same time imposing certain types of limitations on an otherwise unrestricted occupation.

In many cases governments have translated these efforts into a national development programme outlining those areas (systems and species, etc.) which can sustain further growth. However, while these programmes consider various subjects in detail, such as sites for development, pollution, fish diseases, subsidies, etc., only rarely are marketing problems and the demand side of the sector included. Specialised market studies have, however, recently appeared.

No countries have established, or intend to set up, special agencies solely for aquaculture. Rather, units within the ministries for fisheries or agriculture are the central coordinators, exercising extensive consultations with related ministries, such as rural development, resources, and environment, and relevant departments, such as a veterinary agency.

The various types and magnitude of investment and aid made available by governments suggest that many countries attach importance to a sustained development of the sector. The objectives of these national policies are to provide sound employment opportunities in areas where alternatives are few, to provide a production of fish supplementing catches from capture fisheries which are often in decline due to overfishing, and to provide a production sufficiently large for exports to earn foreign exchange.

Countries attach much importance to the development of an aquaculture sector

All Member countries provide some kind of support towards the development of aquaculture and the enhancement of production. In the EEC a coordinated approach has been taken with the institution of a five-year (1987–91) programme; this programme implies a total investment in the sector of some ECU 1.1–1.2 billion in the course of the 5 years. Support is available from the EEC in addition to national aid (mostly in the form of investment grants). Besides direct aid, indirect aid, like research carried out in government-sponsored institutions or government-paid consultants, is available in most Member countries.
Source: OECD, 1989a.

The need for better planning, which takes into account the full range of factors which affect the process of aquaculture development, is increasingly recognized. Some countries, and even regions within countries, have now produced development plans for their aquaculture sectors. These have, however, been few and far between, and they do not amount to a body of knowledge which allows principles to be drawn up specifically for aquaculture sector planning.

Consequently, planning for the sector has to be approached from basic planning principles. This chapter outlines these principles, drawing on different issues and points which have evolved from closely related areas of agriculture planning and, whenever possible, from any recent aquaculture planning.

A sector plan is needed for the optimal and orderly growth of aquaculture, and the efficient use of the country's scarce national resources, so that infrastructure and support services are developed harmoniously. A sector plan is also needed so that aquaculture development leads to positive benefits, while adverse (social and economic) effects are kept to a minimum. Failure to plan properly leads to lost development opportunities, and wasted resources and effort.

The most serious problems relate to the lack of attention given to the management and implementation of the plan, and, in particular, the commitment of both manpower and funds for activities identified by the plan. Sector plans without commitments are worthless, and remain paper exercises which fail to have any influence on the development of the sector.

Aquaculture planning can be conducted according to tried and tested methods which have evolved in planning other sectors. However, given the special nature of aquaculture, and the diverse and localized nature of natural resources required by its production systems and practices, it is important that aquaculture planning is based on an approach which can accommodate such features. Regionally based planning is therefore essential for aquaculture development, avoiding such strategies which emphasize creation of economic growth centres in urban locations.

The ability to conduct planning on a regional basis, of course, depends on the extent to which a country is regionally developed. The objectives of many developing countries emphasize the need for national unity, and government in these cases is normally highly centralized. In reality, therefore, planning in these countries is based on such structures. Moreover, while aquaculture planning is ideally based on relatively large teams covering broad and regionally varying aspects of the sector, developing countries in practice do not have trained resources available to carry this out. Compromises are therefore needed between the ideal and the possible. The practical solution for a developing country is to require that its centrally based planners take strong account of regional needs, rather than conduct a fully fledged regional planning exercise.

Illustrations of current planning activities in aquaculture in this section are largely taken from developed economies. This is unavoidable, given that few, if any, developing countries have yet undertaken comprehensive planning. This bias does not imply that plans for developed countries represent standards which others should attain. On the contrary, good planning requires that a national sector plan reflects true capabilities, potentials, and political realities of that country.

3.1.2 The problems of planning

Planning of all kinds is a serious business. Success or failure can sometimes vitally affect the welfare of many millions of people. Yet planning for aquaculture and agricultural development, which are closely related, has to contend with difficult and special problems. The first section of this chapter attempts to identify these problems.

Success in planning lies in good management and the use of appropriate management techniques. It also requires an understanding of the nature and problems of backwardness specific to any individual region or country. This understanding is often lacking and evident from persistent flaws in the processes of policy. Consequently, problems encountered in seeking better policy directions are not only technical and financial in nature but also political, social, and cultural. These last three, in fact, may be the most intractable.

Sector planning has often been a failure for several reasons. First, there is frequently a wide gap between planning carried out by producers (farmers) and fishermen, and that by government agencies. Second, even if national or regional plans are relatively sound, results may be poor because planners have included only what must be done to achieve the various targets, without showing in sufficient detail how, by whom, or when things will be done. Third, contingency planning is often neglected. Planners fail in all these ways partly because of weak systems of communication between them and the technical ministries, departments, and agencies; between national, regional and local government; and between public and private sectors. Moreover the producers, both farmers and (especially) fishermen, are not usually treated as equal partners in the whole planning process and are poorly informed about important development decisions.

The shortage of skilled managers, specialists, and others, especially at regional and district levels, is an important factor in all these problems of weak communication. The lack of suitable administrative procedures and institutional organization (at all levels) to manage development only adds to the serious gap between plan and performance. Thus actual expenditure on projects in many poor countries often lags far behind official com-

Plate 4: Planning frequently fails because of weak communication between the producers and the government agencies. Here, farmers and extension agents in Nepal overcome this problem in an informal session on a community development project, financed by UNDP and executed by FAO.
Photo: P. Larsen, Courtesy of FAO.

mitments of loans and grants, leading inevitably to failed targets and frustration as project and plan schedules become almost worthless.

Probably the most experienced practitioner in the art of planning is the World Bank. Its long and costly experience has provided valuable insights (relatively well documented) into the whole subject of planning agriculture. Its practices include:

- concentrating on a few selected policy issues and programmes in place of elaborate blueprints;
- effective procedures for consulting and coordination within government departments, and with the wider agricultural and rural community;
- improving the capacity of public sector entities to formulate sound investment programmes – with special attention to the appraisal of major projects; and
- greater flexibility in policy decisions to take account of changing circumstances.

Invariably political issues have also to be faced in the planning process, such as overcoming vested interests which particular policies may create (for example, duties and quotas). These aspects of planning are also discussed in this section.

A sector plan should be prepared within the general framework of a country's national plan (especially in developing countries), but few are. Nevertheless implicit assumptions about the past, present, and probable future performance of the national economy, and its relationship to agriculture (and aquaculture) development, are generally part of the fabric of sector plans. Nonetheless, good planning confined mainly to one sector can still provide some useful results. Unfortunately many sector plans do not have clearly defined objectives or explain in sufficient detail how they intend to reach their objectives or targets.

The time horizon of planning is particularly important and this must always be taken fully into account. Typically, a sector plan has four principal parts (see Figure 3.1), each with its own time frame.

There are endless variations in the scope of sector plans reflecting wide differences in development, political philosophy, and competence. There are also several levels of planning involving a variety of trained people.

1. **An annual or operational plan,**
2. **A medium-term** (5-years) **plan** and, sometimes,
3. **A perspective** (15–20 year) **plan**.

For continuity it is also useful to have some form of:

4. **Rolling plan**

Figure 3.1 The four components of a complete sector plan.

- Choosing objectives,
- Undertaking a stocktaking and diagnostic survey,
- Making demand and supply projections,
- Setting targets and allocating resources
- Choosing strategies and policies, programmes and projects,
- Identifying public expenditure programmes and the financing,
- Identifying institutional changes involved,
- Monitoring, reporting, and control.
- Training may also be included to improve levels of management and decision-making, and to fill the gaps in specialist technical positions

Figure 3.2 The principal activity areas of a sector plan.

Thus it is useful to identify the main features of the sector using an appropriate system of classification. This may cause some problems of definition initially, but the system of classification adopted should be influenced by the structure of a country's agriculture, and/or aquaculture, its level of *per caput* income, and its political and social objectives. Furthermore, to be realistic, the plan should recognize the varying ecological systems of different areas. This provides the basis for regional planning.

In the planning processes, too much attention is often given to plan formulation. In some respects this is understandable as plan implementation, the next stage, is more difficult; but the lack of attention given to implementation lessens the chances of a plan succeeding.

A plan can be divided into its individual components by areas of activity (Figure 3.2). There is a dynamic sequence in which all these plan components should be undertaken. A tentative plan gradually takes shape through iteration during these activities or a process of successive approximation.

Planning within an overall framework

Policy analysts appreciate the link between national and sector planning as they take account in their work of such elements as fiscal policies, monetary exchange rates, and wage and trade policies. However, it is difficult to find a national development plan with a programme for agriculture (and more specifically aquaculture) which is carefully integrated with other sector programmes, or with the general framework of the plan. In some cases many countries prepare sector programmes without any reference to the national development plan.

Most professionals in sector planning dislike programming prepared outside the framework of the national plan, as has sometimes been necessary. They take the position that programming must be done within 'the framework of development planning for the whole economy', or some similar reasoning. The arguments for this point of view are forceful. Development in agriculture and other sectors is clearly interdependent and mutually supporting. Some balance is required in inter-sector growth if allocation of scarce resources between sectors is to be accomplished efficiently, and if the goals of agriculture (and aquaculture) are to conform to national goals.

These arguments are sufficiently compelling for most developing countries to include their agricultural programmes in national development plans. But, unhappily, because national plans are frequently ignored to a greater or lesser extent by budgetary authorities, technical ministries, departments, and autonomous agencies, these programmes are usually not carried out. Moreover, whenever countries fail to meet plan targets, the greatest shortfalls usually occur in the agriculture sector. (Aquaculture has a fairly good record in this respect.) This has happened not only in developing countries, such as India and Pakistan, but also in more advanced and better-off countries, such as the newly independent states of the former USSR and in Eastern Europe.

In poor countries where agriculture dominates the economy, their failure to attain targets largely explains their failure to grow in accordance with general plan targets. Thus, not surprisingly, where the annual rate of increase in agricultural output is low, the corresponding rate of national economic growth is also weak, and vice versa.

Obviously one solution to this dilemma is to improve the whole process of national planning but, thus far, the problems involved have proved to be too difficult and intractable for most poor countries to solve. Results achieved from overall planning have frequently been so discouraging that some authorities contend there is a 'crisis in planning'. On the contrary, the problem may well be 'how to plan in crisis'; in other words, how to plan in times of increased political instability and economic uncertainty, when governments may not be able to exert the required discipline to carry through their medium-term comprehensive plans. It is not necessary here to resolve this issue, but its effect on sector planning cannot be ignored.

In countries where there are no national plans, or where those which exist are not being implemented effectively, the question arises whether it is possible, and/or desirable, to prepare a sector plan (for aquaculture) outside an overall framework. Some authorities contend that even when no general framework exists, any sector plan must be based, implicitly if not openly, on assumptions regarding the performance of the whole economy, with parameters for the expected growth of population, national production, consumption, foreign trade, industry, and so on. Unfortunately, apart from the fact that the quality of the methodology of forecasting economic change for periods as long as 10–15 years is still uncertain, political instability and economic hazards often make short shrift of such projections.

Planning outside an overall framework

There is no doubt that optimal allocations of resources to economic sectors, and to programmes and projects within each sector, can only be attempted within the framework of a national plan for the whole economy (and yet the record so far is not always encouraging). Many professionals in sector planning are willing, however, to forgo this ideal as they work outside an aggregate framework, either where existing national plans are not operational or where it is considered impractical to prepare national plans which have little chance of implementation. Moreover, experience shows that, however undesirable in theory, in practice it is possible to achieve considerable success where planning is confined to agriculture sectors, which will include aquaculture. For example, the World Bank, sometimes in cooperation with FAO, has prepared and is preparing agriculture sector plans in a number of developing countries outside a general planning framework.

3.1.3 Aspects of sector planning

Objectives of planning

Probably most countries have carried out some planning activities for aquaculture, whether as part of a national development plan or not. Such planning varies considerably from one country to another because there is no widely accepted set of principles for ordering aquaculture development. If it is accepted that a sector plan starts with one or more attainable objectives, indicates the resources available to achieve them, explains how these resources are to be distributed among specific projects, and states the strategies, policies, and institutional arrangements to be used, then few plans strictly meet these criteria.

Many sector plans do not have clearly defined objectives, unless they are to carry out a bundle of projects which are often unrelated to each other. This omission is more rooted in politics than in the inexperience or incompetence of planners or a minister, and few of the projects which may make up a plan are carefully prepared and complete with estimates for their 'cost-effectiveness'. Even where objectives of the plan are clear, the means for achieving them may not be spelled out in detail simply because the required analytical work has not been done.

Some plans do not indicate what resources are to be allocated to each project in the programme, which may be because there are not enough well-prepared projects. And if resources are in fact allocated to a sector only in aggregate, there is no knowing whether the amount is sufficient or not. Only when projects have been prepared in some detail, and their location determined, is it possible to estimate their cost with reasonable accuracy, together with an appropriate estimation of factors for inflation. Another problem is that plans which try to include all public investment in aquaculture concentrate only on two or three sub-sectors, such as processing and cold storage, seed stocks, and research. Finally, plans which claim to be comprehensive often exclude a large part of private investment. This is often due to the difficulty of collecting these data.

Many plans try to achieve impossible objectives by including a long list of projects which cannot all be completed within the plan period. For example, it is quite common for even small countries to be over-ambitious in listing activities to be carried out in a five-year period.

Many five-year plans are over-ambitious

A National Plan listed the following projects to be completed in five years:

- about 20 on-growing units for euryhaline species, with total production capacity of about 4000 t,
- 10 complete units for the production of euryhaline species, including hatcheries with a 1 million smolt capacity,
- 3 hatcheries with on-growing units,
- 6 shrimp production units of freshwater and seawater,
- 20 oyster culture units of 2000 t capacity,
- 15 eel culture units of 1200 t capacity,
- 3 crayfish culture units of 30 t capacity,
- at least two salmon culture units of 80 t capacity,
- 7 units of various species of fish and crustaceans with marketing importance.

The practice of producing long and detailed lists of objectives is defensible if the planners recognize that some projects will give little or no return in the plan period. But if there is no distinction between such projects and those which are expected to yield outputs and income in the period of the plan, then it is obvious that the planners do not have a clear idea of the limitations of the time horizon of the plan.

Planners

A wide variety of men and women of different disciplines, backgrounds, skills, and training take part in planning. Planning should be, and often is, a team effort operating at several levels, namely at headquarters of the ministry or department of planning, within any special planning commission or unit (depending on its official status and influence), and at regional, district, and local levels.

An illustration of a regional planning team in a developed country

'The Cook Inlet (Alaska, USA) Regional Planning Team is made up of six voting members, three representing the Cook Inlet Aquaculture Association (CIAA) and three representing the State of Alaska Department of Fish and Game (ADF&G). In the case of the latter three there is one representative each from the Fisheries Rehabilitation, Enhancement and Development Division, the Commercial Fish Division, and the Sport Fish Division. In addition the non-voting position of Chairman is currently held by a representative of ADF&G.'
Source: Alaska Department of Fish and Game (USA), 1982.

In this example members were appointed to this team from the Cook Inlet Aquaculture Association shortly after its official formation, and have been active in the planning process since 1977. In 1979 additional support was sought for the team's activities, and a consultant was retained to aid in putting the plan together. Later that consulting function was supplanted by a planner attached to the team itself.

The Cook Inlet Regional Salmon Enhancement Plan is limited in its scope, namely it was a partial plan for dealing only with Pacific salmon and with a particular region of Alaska (Cook Inlet) because of its special resource. However, it was still prepared by a regional team of fisheries sector administrators, sector managers, and industry representatives, supported by a consultant (in reality a consulting company of interdisciplinary specialists, engineers, and economists), and eventually by a full-time planner to monitor, evaluate, and propose any changes.

In developing countries, especially those with little or no aquaculture industry, there will not be resources available for the detailed coverage in the planning effort illustrated by the example. The size of the team will have to be tailored to what is possible within the country's resources.

A second distinction can be made between those working on a national plan and those involved in preparing, implementing, and evaluating programmes and projects which are parts of that plan. Thus there is a wide difference in job requirements between, say, a high level planner working at government headquarters on an equal footing with others in the same staff category, such as staff in the prime minister's office, the treasury, or other key government offices, and extension workers who also play a vital role in the planning process. A further distinction can be made between the small number of individuals who actually put a plan or project together and those who collect the information without which any plan is incomplete, or even useless.

The success of planning depends in large measure on whether the main features of the plan are understood and accepted by the community it affects. An organization, such as an Aquaculture Coordinating Council, can play a useful role in this regard, serviced by planning officials at all levels of administration. For example, even within a small aquaculture planning project, such as the preparation of culture-based inland fishery catch projections in a five-year plan period, a considerable number of individuals are likely to be involved. The nucleus of 'planners' will include policy analysts, planning economists, commodity, trade, aquaculture and fishery specialists, statisticians, econometricians, and advisers on aquaculture and fisheries management, finance, and administration.

At times some staffing functions may be combined, where there may be staff limitations, lack of data, or to save time and money. The planning unit begins its work of assembling existing data and documents relevant to stock recruitment, fish catches, and productivity. This is followed by consultations with other government departments, organizations, agencies, and individuals who, because of their functions and responsibilities, have special knowledge of, and/or are likely to represent important points of view on specific aspects of fish production and outlook.

If external financing agencies are already, or are likely to become involved, then they are also consulted. It is here that specialists at the local level, such as extension workers and others involved in the sector, play a useful role because of their working knowledge of fishing communities who must be closely involved in the planning work if it is to have real commitment. Once all the data are collected, evaluated, and analysed, preliminary output projections can then be prepared which are almost certain to be adjusted later as the plan is put together.

Time horizons of plans

Because sector plans of most countries are part of their medium-term national plans, ranging from three to seven years, with a mode of five, the time horizon of such plans is generally too short for bringing about basic changes in a given sector. Therefore it is essential in planning for aquaculture development to take a longer than medium-term look into the future. How far ahead depends on the main changes desired.

Most planners in aquaculture adopt a perspective of a decade or more. Unfortunately a 10–15 year period is too long for the formulation of plans of action for aquaculture development. Apart from the shortage of basic data required for long-term planning, so many unforeseeable changes in domestic and external conditions are likely to occur that a detailed plan for 10–15 years ahead is almost certain to be outdated very quickly. Consequently, planners recommend that a long-term or 'perspective' plan for 15 years or more provides only broad lines of desirable sector development.

A perspective plan is not couched in precise terms but only formulates the strategy of sector development in terms of the country's needs and technical possibilities in particular fields, such as aquaculture, forestry, land development, and so on. With such a perspective it is possible to define the general direction in which long-term development should take place, in accordance with government's objectives.

The strategy can then be supplemented with projects and policies for achieving development objectives for aquaculture, consistent with national development goals. However, these projects and policies need not, and indeed should not, be incorporated back into the perspective sector plan because the time horizon, as noted, is so long that it will undoubtedly be revised, perhaps several times. For this reason it is generally recommended that the medium-term plan for the development of aquaculture is spelled out by a series of medium-term projects, with details of projects extending beyond the medium-term to be included in future medium-term plans. While each project needs to be studied and prepared in detail, the timing and scale of its inclusion into a medium-term plan depends ultimately on unforeseeable factors which may require it to be held back, brought forward, or even cancelled.

It is extremely important that both the perspective and medium-term plans are reviewed periodically, and adjusted to take account of changes brought about by domestic and external events. A review of the perspective plan can be made every few years. For medium-term plans the review is usually an annual exercise, and may be done in one of several ways:

- Revise the entire medium-term sector plan every year for the remainder of the plan period. This is both difficult and time consuming because it requires an almost full-time commitment of trained personnel which few countries possess.
- Prepare annual plans which take account of progress made to date, allocate available resources to projects as a basis for budgetary appropriations, and set out specific instruments of policy and other measures for achieving development objectives. While this is an easier way of adjusting to unforeseen changes, it has one serious defect. Year-by-year revised annual plans deviate substantially from the medium-term plan so that it no longer provides much guidance for their preparation.
- Introduce, to avoid the difficulty noted above, a so-called 'rolling plan'. Here an additional year is added to the length of the medium-term plan at the end of each year, after making appropriate adjustments due to changed circumstances in the year just passed. The medium-term plan thus provides a look ahead at the end of each year for an undiminished period. This approach has the added advantage of providing a medium-term plan with up-to-date benchmarks for annual plans, which have to be prepared anyway to make the medium-term plan operational. If the 'rolling forward' technique is also applied every few years to the perspective plan, it too can be updated to provide guidelines for rolling the medium-term programme forward each year.

The use of perspective, medium-term, and annual plans is an ideal solution to the time-horizon problem. The perspective plan permits looking far enough ahead to identify in broad outline the main directions of sector development. The medium-term plan spells out in greater detail than the perspective plan interim goals which must be achieved in the medium-term to attain longer-term targets. And the annual plans make the medium-term plan operational by:

- reducing medium-term targets (which are usually stated in terms of the end of the medium-term plan period) to annual targets to allow allocations required for their achievement to be incorporated in annual budgets, and
- enumerating in sufficient detail the measures to be adopted to achieve the plan's objectives.

Thus the plans fit closely together.

This approach to sector planning is largely associated with conventional macro-economic planning, which most authorities now agree requires preparation of overall perspective, medium-term, and annual development plans. The difficulty is that, in practice, few low-income countries have succeeded in formulating the three types of development plans, and in integrating annual plans, where they exist, with annual budgets. Although some countries have, or have had, perspective development plans and almost all have, or have had, medium-term plans, few have made these plans operational with annual plans closely linked with government budgets.

In practice, therefore, sector planning in developing countries has been largely limited to the preparation of only one medium-term plan, the duration of which has varied. Plans of shorter duration, usually a year, are also frequently used to translate longer-term perspective plans into action. This combination has proved to be workable and may be the practical solution for aquaculture planning in developing countries.

On the basis of this experience, it would seem acceptable to prepare only one medium-term plan and an annual operational plan. While the longer-term perspective plan could cover a period of only five years in countries which have few projects with long gestation periods, many countries are likely to find perspective plans for 10–20 years more suited to their needs.

In countries where talent and time exist, a rolling plan for the first three years of the multi-annual plan (more detailed than the rest of the plan but less detailed than the annual plan), could be given some priority because it enables adjustments to be made as required by changed circumstances. In addition, the multi-annual or medium-term plan as a whole might be rolled forward every few years.

The scope of plans

The scope of sector plans reveals as much variation as their duration. Many countries prepare what they term comprehensive sector plans for agriculture but this does not necessarily mean that each sub-sector receives equal attention. Plans for forestry and, even more, fisheries are likely to be less detailed than for crops and livestock. Nor does private investment always receive the same emphasis as public investment. At the other extreme, some plans make no attempt at a wide coverage and limit the document to only one sub-sector, or even one crop, such as the Region Salmon Enhancement Plan for Cook Inlet, Alaska.

Where possible it is desirable to deal with the whole sector, but limitations of planning data, training manpower, finance, and time may make partial coverage inevitable. Although partial planning is not as good as comprehensive planning in theory, it sometimes works better in practice, particularly where one or two sub-sectors dominate the scene. Partial planning should be avoided, however, where the government is attempting to diversify agriculture or aquaculture and to shift the emphasis away from one or two commodities. Even so it is well to be selective in emphasizing one sub-sector over another.

Classification of systems for the sector

For emphasizing a particular sub-sector, as well as for other reasons, a suitable definition of the aquaculture sector is essential, which prescribes its limits and classifies or divides it into appropriate components. There are four ways in which this sector has generally been defined, namely:

- by employment,
- by size and type of holding (or business),
- by type of economic activity, and
- by output.

For classification by *employment*, the general concepts, definitions, and classifications of labour force, employment, unemployment, underemployment, etc., as recommended by the International Labour Office, are all elaborated in the framework of an industrial society on the basis of the existence of well-organized establishments with employers, workers, fixed wages, fixed hours of work, etc. They are difficult to apply to aquaculture, where business is often done on a family household, or an informal basis.

The amount of aquaculture work to be carried out is irregular and may change from day to day. It is also seasonal. There are times when the work is intensive, and others when it is almost non-existent. For these reasons, the statistician is not so much interested in measuring the volume of employment and unemployment on a fixed date, or during a short period of time. He or she is much more interested in quantifying the total volume of work to be carried out on the holding, or in the business, its distribution according to source of labour inputs (family, non-family, regular, part-time or spare time, etc.), and its relationship with size, type, and other characteristics of the holding.

The concept of time worked in aquaculture is much more difficult to define than in most other fields of activity. Such work includes working on production, preparing produce for and taking it to market, buying and transporting inputs, record keeping, etc. There is no fixed working place. Distances between different sites of work may be long and time-consuming. Care is thus needed to include all the relevant periods of work and travel when recording the time worked by the producer, family, and paid workers.

A useful concept is the 'man-year' unit of work to take account of the wide variety of labour used. This unit is defined as the amount of aquaculture work (including farming, if appropriate), carried out by the able man during a year, the year being defined as a number of hours (e.g. 2200 hours) or days (e.g. 275 days). Appropriate adjustments are made for the labour of older workers, women and children.

This kind of analysis requires data and expertise, however, which are not always available. To avoid difficulties, any planning exercise should explicitly state how the labour force has been counted.

The *size and type of holdings* (or business) are major characteristics investigated in agricultural or aquacultural censuses. The concept of size of holding can be defined in a number of ways according to the basis of measurement. It can be measured in terms of area, output, labour, etc.

In censuses of agriculture, especially those carried out on a complete enumeration basis, practical considerations make it necessary to limit the enumeration to holdings or businesses above a certain limit of size. Further, most of the agronomic surveys carried out, subsequent to a census, use as a frame the results of the census of agriculture and thus do not include holdings below the low limit. This results in a situation in which the most vulnerable section of the agricultural population (including fishermen) may not be investigated at all. To remedy this situation, planners should specify that some key questions be introduced in the censuses of population to identify these small holdings or units which could then be investigated on a sampling basis. Planners will also find it helpful if holdings or businesses are divided into the following three sectors:

- the *private sector* which includes all holdings operated by individuals or households;
- the *public sector* which covers those holdings operated by a central or local government, either directly or through a special body;
- the *collective sector* which covers the holdings operated by a group of persons who, voluntarily, or by mandate of the governing authority, join together to exercise land rights in common. However, if a member of a collective receives a plot for his personal use, this plot will be part of the private sector.

Another classification of holdings is based on the degree of utilization of advanced techniques where holdings are sub-divided into modern, progressive, and traditional. However, the definitions generally used are rather vague and often differently interpreted by individual countries.

Defining the agricultural industry by the *type of economic activity* raises the problem of how to identify establishments which are engaged in a variety of agricultural enterprises, e.g. crop production and aquaculture. The two criteria set by the UN International Standard Industrial Classification (ISIC) to deal with this problem provide that:

- the farming units included in a given activity group must account for the bulk of the output of these units, and
- the group must contain most of the farming units which produce most of the class of foods and services which characterize the activity concerned.

The best known example of a classification system which groups farming (and other) establishments by their economic activity is the abovementioned UN system. The ISIC arranges economic activities into four categories: (a) major divisions, (b) divisions, (c) major groups, and (d) groups.

The UN International Standard Industrial Classification identifies establishments engaged in agricultural enterprises

There are ten major divisions in the ISIC of which major division 1 covers agriculture, hunting, forestry, and fishing. This major division is further sub-divided as follows:

Division 11	Agriculture and hunting		
	111	1110	Agricultural and livestock production
	112	1120	Agricultural services
	113	1130	Hunting, trapping, and game propagation
Division 12	Forestry and logging		
	121	1210	Forestry
	122	1220	Logging
Division 13		130	Fishing
		1301	Ocean and coastal fishing
		1320	Fishing not otherwise classified (includes aquaculture)

The most common way of defining the aquaculture sector is by the *output* of its various commodities: e.g. finfishes, crustaceans, molluscs, seaweeds, other. Examples of current output classification are given in the current FAO Fisheries Department Aquaculture Production Statistics Form (see Figure 3.3), the FAO Aquaculture Production Reports, and in Country Reports prepared by the Office of Economic and Cooperative Development (OECD). However, these classifications have been used primarily with international comparisons in mind, and are not necessarily suitable for

FAO/CWP
FORM FOR REPORTING STATISTICS ON AQUACULTURE

FISHSTAT **AQ**

(a) YEAR	(b) COUNTRY OR AREA	(c)	(d)	(e)
19		Form for reporting details of type of culture. Do not record quantity produced on this form.	For culture in brackish or marine waters, please specify FAO fishing area	SHEET No 1

A	B	C	D	E	F	G	H	I	J	K	L	(A)
Line		METHOD OF CULTURE (by main categories of aquatic organisms)	Freshwater culture			Brackishwater culture *			Mariculture			Line
			No of Units **	M^3 in '000	Hectares	No of Units **	M^3 in '000	Hectares	No of Units **	M^3 in '000	Hectares	
1		**FISHES**										1
2		Ponds and tanks										2
3												3
4		Enclosures and pens										4
5												5
6		Cages										6
7												7
8		Raceways and silos										8
9												9
10		Barrages										10
11												11
12												12
13												13
14												14
15		Others										15
16												16
17												17
18												18
19		**CRUSTACEANS**										19
20												20
21		Ponds and tanks										21
22												22
23												23
24		Enclosures and pens										24
25												25
26												26
27		Others										27
28												28
29												29
30												30
31												31
32												32
33												33
34												34
35												35
36		**MOLLUSCS**										36
37												37
38		On bottom										38
39												39
40												40
41		Off bottom										41
42												42
43												43
44		Others										44
45												45
46												46
47												47
48												48
49												49
50												50
51												51
52												52
53		**SEAWEEDS**										53
54												54
55												55
56												56
57												57

		HATCHERY/NURSERY OUTPUT		Number of larvae, fingerlings, juveniles, etc. released, i.e., used for stocking			
				Number released			
		Local Name	Scientific Name	To a controlled environment	To the wild	Total	
58		**SPECIES**					58
59							59
60							60
61							61
62							62
63							63
64							64
65							65
66							66
67							67
68							68
69							69
70							70
71							71
72							72
73							73
74							74

* See Section 4 of the instructions
** See Section 5 of the instructions

NOTE Shaded areas represent concepts of no technical validity and consequently figure should not be inserted into these spaces

FID 21 · 3.93 · U9915

Figure 3.3 Classification of the sector by outputs: The Aquaculture Production Statistical Data Sheet (Source: FAO).

sector planning within particular countries. France, for example, collects and reports production of its active aquaculture sector as and together with its agriculture data.

Which classification system to use

The classification system eventually adopted should be determined by the structure of a country's aquaculture, its stage of development, and national objectives. Each country must adapt the system to its own situation. Thus, where impact of a development plan for aquaculture on the balance of payments is important, it should be possible to distinguish between export- and domestic-oriented activities. If the main effect is a general increase in output and international comparisons are not important, then a commodity classification may be adequate. However, if there is an interest in income distribution, or levels of living of specified groups of fishermen or farmer-fishermen, then a system based on economic activity is needed because it will provide for units to be classified according to size, character, technology, and production organization (besides facilitating international comparison).

As indicated earlier, the ISIC system not only provides for productive units of different kinds, technology, and organization to be identified, it also permits classification by output of commodities. A system based on economic activity, like the ISIC, can therefore meet planning requirements where objectives include the improvement of particular kinds of business, increased output, or both.

The need to select or design a classification system which is appropriate to the particular country gives rise to a wide variety of solutions. In a developing country, for example, where aquaculture activities are limited by technical factors to (say) culture-based fisheries involving the production of one species, an elaborate classification of output or economic activity is not required; there is only one category of output, and aquaculture activity is mostly limited to hatching and seed supply. In such a case, the main need is for information about the requirements for trained manpower (skilled hatchery personnel) and cost benefit assessments, and planning for the sector will depend on this detailed information being available. The classification system is therefore designed with this in mind.

Similarly, in a sector as currently exists in Norway, where aquaculture is mostly based on the production of a single species (Atlantic salmon), on a single set of universally applied production methods (cage farming), and where production holdings are similar in size (volume maximum 12 000 m^3), then elaborate classifications of output and holdings are redundant. On the other hand, where there are pressures to maximize social benefits by adopting a strategy to develop a given preference to holdings managed by families or local cooperatives (as exists for salmon farming in Sweden and Finland) then the system of classification is designed to divide the data between these categories of holdings. Otherwise efforts to develop this strategy will be badly handicapped.

More complex aquaculture sectors demand more complex classification systems. In a country such as Greece, for example, the small but potentially useful sector already has a wide range of production practices, species under culture, environments where aquaculture is exploited, and forms of management. Salmonids, carps, mussels, eels, and marine fishes, are all produced through a variety of practices, in various environments, and under different management arrangements. Moreover the sector has some tradition, dominated by extensive lagoon and freshwater production, and there are interactions between these and capture fisheries. Established legislation limits access to these fisheries to local groups, with the effect of inhibiting or preventing the development of privately-managed aquaculture production in these areas.

Such complications are acute in parts of Greece but, in this respect, the country is similar to many developed and developing countries (such as Mexico). Traditional practices, culturally-determined attitudes, and existing social structures all create conditions which aquaculture planners cannot ignore. Many sector plans, unhappily, do neglect these features. Such plans either fail, or lead to social disruptions and conflicts.

It follows, therefore, that data must be arranged in such a way that complications of the type experienced in Greece are understood and accommodated. In Greece an elaborate classification of aquaculture is certainly needed. Planners, politicians, and sector managers therefore require information on returns on investment and levels of employment in the traditional extensive lagoon culture and modern intensive cage culture to make comparisons, and then choose how the government should intervene in the sector. If these choices are to be informed, and the interventions effective, the classification system has to be designed to provide relevant and detailed data to enable this analysis to be made.

In summary, the classification is adapted to the particular situation which exists, or which is potentially feasible, in each country. However, much depends on the quality and coverage of a country's statistical services in deciding which classification system to choose. Ideally statistics should conform to the needs of planning, rather than the other way around. In aquaculture, at the present time, this is not possible. Therefore best use has to be made of whatever data are available, however frustrating this may be. In developing countries it is necessary to adopt a pragmatic approach to this problem if any form of planning is to be established.

Finally the classification system provides a means of producing a matrix of data for aquaculture. The next stage is to analyse the data, and the planner requires a framework for this analysis. In the last few years, new frameworks for aquaculture include the approach (as outlined in Chapter 1), where a sector is divided into its different forms of organization and management existing within it, such as small-scale or vertically integrated aquatic farming, culture-based fisheries enhancement, etc. Another frame-

1. **Consumption of fish and shellfish and the regional markets**
   - Benefits of fish and shellfish to the consumer;
   - Consumption characteristics in the region;
   - Annual consumption of fish and shellfish (in region, with comparisons, and by country, if variable or atypical);
   - Gross market data (regional/national statistics, GNP, population);
   - Specific market data (marketing characteristics, distribution, associations, popular species and prices, trends, subsidies, product forms, exports, value, trade tariffs);
   - Information for the trade (consumer publications, market publications, market trials);
   - Technical assistance projects in the sub-sector.
2. **Aquaculture production of fish and shellfish in the region**
   - Characteristics of production in the region;
   - Regional production data (gross production for regional comparisons, national production data, production by genera and/or species);
   - Production systems and practices in the region (extensive, semi-intensive, intensive systems and practices, areas under production);
   - Producers in the region (numbers of producers, distribution characteristics);
   - Organizations of producers (cooperatives, NGOs, associations);
   - Financial investment by public and private enterprises;
   - Technical assistance projects in the sub-sector;
   - Capital assistance projects in the sub-sector.
3. **Local infrastructure**
   - Extension services in the region (characteristics of national organizations, agent qualifications, career advancement, agent complement, resources in facilities, transportation, methodologies of service, extension materials);
   - Training of extension agents (vocational schools, transfers, special training courses);
   - Seed production facilities in the region (hatcheries, field stations, production, species, organizations of seed producers, cooperatives, NGOs, subsidies, characteristics of trade);
   - Manufacture of feed and fertilizers in the region (manufacturers, millers, distributors, products, costs, organizations of suppliers, cooperatives, NGOs, subsidies, characteristics of trade);
   - Manufacture of equipment for the industry (engineering contractors, fabricators, suppliers, organizations of trade, associations, trade shows, equipment subsidies, characteristics of trade);
   - Other services for the industry (consultants in disease diagnostics, water chemistry, programme and design engineers);
   - Local credit programmes (institutions, volume, terms, conditions);
   - Trade publications for producers;
   - Technical assistance projects in the sub-sector.
4. **National infrastructure**
   - Manpower training in aquaculture technologies (characteristics of background education, undergraduate training and institutions, facilities, post-graduate training and institutions, facilities, special technical courses);
   - Manpower training in management (business, farm, credit, research);
   - Aquaculture research in the region (institutions, programme objectives, staff complement, laboratories, outdoor facilities);
   - Management of credit (credit institutions, policy, programmes);
   - Associations of professionals in the region (societies, networks, meetings);
   - Information for professionals (international publications, regional and national publications, data bases);
   - Technical assistance projects in the sub-sector;
   - Capital assistance projects in the sub-sector.
5. **National sector management**
   - Government policy in the region (existing statements, individual or *vis-à-vis* other sectors, social objectives, health, welfare, employment, women, existing national plans and budgets for aquaculture development);
   - Government organization in the region (characteristics *vis-a-vis* agriculture, fisheries, other, staffing policy, staff complement, responsibilities);
   - Aquaculture legislation (licences, permits, taxes, subsidies, environmental regulations for species transfer, quarantine, genetic conservation);
   - Trade and business associations;
   - Information resources for management;
   - Technical assistance projects in the sub-sector.
6. **Regional sector management**
   - United Nations organizations in the region (member countries, general objectives relevant to the sector, history of support, financial investment);
   - Other organizations in the region (member countries, general objectives relevant to the sector, history of support, financial investment);
   - Economic communities in the region (member countries, general objectives relevant to the sector, special programme activities in the industry);
   - Regional and global aquaculture development projects.

Figure 3.4 A sector framework for data collection (Source: ADCP/REP/87/27, FAO, Rome).

work is illustrated by the example in Figure 3.4. This structure was used by the ADCP in its ten regional aquaculture sector surveys, and in all its planning documents. It was also used by ADCP to survey the employment component of the aquaculture sectors in Norway, and several other countries. It uses a framework which divides the sector into six levels, with principal components for the target groups (the consumers and the producers), followed by four components which support them, namely local infrastructure, national infrastructure, national sector management, and regional management.

3.2 Regional planning

3.2.1 The need for regional planning

A national aquaculture plan cannot be realistic unless it takes adequate account of the varying climatic and ecological factors in the different areas of the country, specifically temperature, rainfall, humidity, soil and water resources, drainage, topography, and altitude. These all condition the methods used to increase production. For example, Greece and Turkey are countries where there is a high degree of variation between regions (mountainous interiors and extensive Mediterranean coastlines), which influence the potential for a variety of aquaculture products and methods of production.

Different regions have varying resources and potentials

Turkey has aquaculture potential in the Black Sea, the Aegean, and on the Mediterranean coast, and in extensive freshwater areas inland. Planning requirements for each are all different through geography, indigenous aquatic species, and economic development.

Greece already has regional variations in its national production, for example: percentage shares of production by region based on the number of farms.

Region	Trout	Carp	Mussels
Thrace	1	18	
Macedonia	25	30	78
Thessalia	3	18	
Sterea	11	7	22
Ipiros	53	22	
Peloponnese	6	5	
Crete	1		

Source: MacPherson, 1987.

Some regions are potentially more productive than others. Those with easy access to large urban markets have had a better start, while others in remote rural areas lag behind. If disparities between regions are to be reduced, then the needs of the poorer regions will increase. Consequently, each region is likely to have its own development requirements, and within each region there may well be considerable diversity.

Regional development may be quite independent

'State and local government. It is difficult for both US and foreign observers of the political scene in Washington DC to comprehend that the same decisions over aquaculture are being indecisively debated at just about every one of the 50 state capitals around the nation, and, in some special cases at municipal government levels as well. Fortunately for aquaculture's progress, at the state level it has been possible to deal with specifics more easily, since there are fewer or weaker special interest groups. Perhaps the leader among states in moving towards positive legislation for aquaculture is the island state of Hawaii. In a period of five years it has drawn up major state documents reviewing its resources, and the inhibitions and regulations which impede developments, and providing a comprehensive plan for the future. The state government has backed its plan with funds and set out boldly on its own independent path.'

Source: Nash, 1979.

Ideally a national aquaculture plan should include plans for all the regions. This is a difficult task, especially if the national plan is attempting to deal with this industry in a comprehensive way. This situation explains why many plans are usually couched in broad terms, with few details given about the problems and needs of specific areas.

Within recent years, however, countries with and without national plans have been turning increasingly to regional planning. This has partly come about from growing political awareness in the less favoured regions of the vital role central government plays in the allocation of investment, and of their consequent need for organizations in the poorer regions to exercise more pressure on the central authorities. Paradoxically, national and sector planning has encouraged this political development, as one result of such central planning is that well-endowed regions often get the lion's share of infrastructure and other substantial investment projects, leading to a widening of regional disparities.

Experience shows that regional sector planning has taken three main forms. The first is a plan consisting of sub-plans for all the regions coordinated in line with national development objectives and strategies. The second covers one or several selected regions within a country. This approach is sometimes adopted by less-developed countries which do not have resources to carry out a comprehensive programme. They prefer to concentrate their effort in one or a few regions where total additional returns are likely to be higher than if the same resources were spread more thinly. The third form is one which is part of a comprehensive plan for a specific region, or where it is considered desirable to develop national resources located in a given region; for example, a river basin for irrigation or other purposes, such as the Aswan Dam Development Authority established in Egypt.

3.2.2 The methodology of regional planning

There is no widely accepted theory which identifies the key elements in regional development. Nevertheless, some or all of the following propositions are generally accepted by regional planners. Regional development:

- largely depends on the extent to which the region's export capabilities expand,
- requires the mobilization of opportunities as they arise, and
- is most effective within the framework of an integrated regional or sub-regional plan.

The argument in favour of developing the export base assumes that a region has a comparative advantage in one or more agricultural or manufactured products or in, say tourism, or in a combination of these sectors which can be expanded by the establishment of specialized marketing organizations, improved credit and transport facilities, a trained labour force, and complementary industries. The need to mobilize government and individuals assumes that ability to recognize, evaluate, and act on investment opportunities, as well as willingness to accept risk and adopt innovations, is frequently lacking in undeveloped regions. And the notion that economic development must take place within an integrated regional or sub-regional system comes from the belief that only then can planners obtain a proper understanding of where development arises, and how it is diffused through the regional or sub-regional economy.

The model which was once most widely accepted among regional planners may be stated in two propositions:

- growth and development are initiated in a limited number of urban 'growth centres' or 'growth poles', and
- are then diffused from these centres to their respective hinterlands by growth-inducing innovations and market mechanisms.

The model necessarily requires the concentration of resources on urban development, at least in the early stages of regional development. Those who hold to the growth centre or pole model contend that theoretical and empirical findings show that regional growth and development take place most effectively through the articulated development of a system of urban centres, arranged in a hierarchy from small agro-urban towns through several intermediate types to the metropolis. Growth impulses are transmitted from the metropolis to the rural areas, with larger centres retaining activities of greater scale and capital-intensity, and smaller centres developing activities which can be performed at lower scale for more local markets, or in which the capital–labour ratio is low; and with all centres diffusing growth to their hinterlands. The city is regarded as the 'engine' of regional and national economy. Agriculture tends to be more efficient, commercial, and prosperous in the vicinity of cities, and becomes less so as the distance from urban centres increases, until in the less accessible marginal areas agriculture (including aquaculture) is characterized by backward, subsistence systems. An essential element of this model is the provision of a good network of roads and transport facilities providing efficient distribution of goods and services required for rural development.

The importance of balance between the centre and the regions

'Memoranda of Understanding (between the Provinces and the Federal Government) to promote commercial development should be entered into on a bilateral basis so as to recognize regional differences (Statement of National Policy Goals for Aquaculture, Canada, 1988). Where regional or provincial governments have less autonomy than countries like Canada, increased regional initiatives can result from efforts to decentralize. For example, in Greece, there are official efforts to decentralize planning and development powers. Law No. 1622/86 defines new arrangements for direct elections to the 2nd tier and provides for increased powers at this level. Law No. 1416/84 gave improved powers to local authorities to set up and participate in development companies and commercial enterprises. Some local authorities are now using these powers for promoting aquaculture development. The development of aquaculture, with its highly localized potentials and problems, should be facilitated by this devolution of economic planning powers.'
Source: MacPherson, 1987.

In many developing countries, the national objectives would require a subordination of regional objectives, and this would affect the aquaculture planning process.

Most countries which have made an attempt at regional development have relied on growth centre strategy, but more recently emphasis is shifting toward rural transformation. Regional planners in these countries have tried through various means to create external economies and linkages at selected urban locations. The usual approach has been through a government-sponsored industrial development plan. Frequently, the plan is directed toward strengthening existing towns and cities which are considered to be strategically located, or on the threshold of sustained growth. In some cases new towns are constructed, especially in areas lacking cities and towns of any significant size. After identifying the locations to be developed as growth centres, a government usually devises a public investment plan and economic incentives for the private sector, designed to guide industrial investment toward these centres. The plan may include provision for direct public investments in industry, industrial estates, other infrastructure and social programmes, as well as low-interest loans, grants, tax incentives, and subsidies to private investors.

The relatively few poor countries which have begun to plan for regional development, and the mixed results they have obtained, tend to show that the state of regional planning is still in its early stages in these countries. Experience of economically advanced countries in dealing with problems of backward regions suggests that attempts to restructure the regional organization of any economy are likely to be extremely costly, and only marginally successful. In less developed countries, where there is as yet little infrastructure and other heavy investment, opportunities exist to structure investment in ways which may prevent, or at least reduce, wide regional inequalities.

Nevertheless it is by no means clear that the way to start development of a region is by making some small town or large village grow rapidly by heavy investment of mainly industrial inputs, along the line advocated by the growth centre theoreticians. In fact, industries may not find it easy to expand in a new urban environment if there has been no prior agricultural growth. In such cases, they may find it difficult to establish linkages with the rural areas, and finish up as isolated out-stations of the modern sector in an unchanged rural environment. Attention is shifting to giving more priority to agricultural and rural development, on the Chinese principle of 'walking on two legs'.

3.2.3 Problems posed by regional planning

Despite its problems, some of which are outlined below, regional planning continues to hold the interest of governments.

Size and definition of a region

The first problem is delineation of the appropriate size and definition of a region. In general, the appropriate size is one which is not so small as to lead to undue fragmentation of national sector planning efforts, yet small enough to enable rapid feedback between farm units and regional planners. In the Philippines, which consists of many islands, the entire island of Mindanao was originally established as a region (it was later sub-divided into four regions); in Italy the southern part of the country below Rome was constituted as one development region; in Colombia it was one single valley which constituted the Cauca Valley Region, and the Gezira Region in Sudan was defined on the basis of irrigation possibilities. Turkey has identified some 12 regions for different development priorities.

It is thus apparent that a region may be defined in different ways. Its boundaries may be determined:

- geographically, e.g. a valley or river basin,
- demographically, for example, an area inhabited by a specific population group, e.g. a nomad group,
- administratively, i.e. one or more political sub-divisions of a country, and
- functionally, e.g. an area within the sphere of influence of a particular urban centre.

Because regional development usually includes a variety of functional activities, which may be as different as the construction of irrigation systems, the expansion of inland fisheries, or programmes for family planning, the most suitable set of boundaries for one purpose may differ very much from those for another. Consequently the definition adopted for a region is likely

Plate 5: Regional planning for development usually requires integration of more than one sector. Project planners discuss the water and agricultural proposals for the Chaletenango Department in El Salvador with a local mayor. This Technical Cooperation Project financed by FAO is for preparation of a Master Plan for agriculture and fishery development.
Photo: L. Dematteis, Courtesy of FAO.

to be a compromise, which is better for achieving some objectives than others. In practice the realities make it necessary for most countries to define regions by their political-administrative boundaries, rather than by what may be considered the more natural boundaries. This is not always bad, as the introduction of a new region adds new 'relay points' into the bureaucratic system and may cause serious administrative delays. For these reasons, even if a major development effort, for example, an irrigation area, has contours which deviate so sharply from existing political-administrative jurisdictions that new regional boundaries seem necessary, the new region should be built up from multiple, or components of, existing jurisdictions.

Definitions of regional needs and objectives

A second problem involves defining the basic social and economic difficulties to be resolved. Everything depends on who tackles this issue. If officials set the objectives and priorities, the results may be (and are likely to be) different to those set by farmers and/or fishermen themselves. Too often planners and other officials look on the farmer as someone to be acted upon instead of to cooperate with. They frequently take a paternalistic, or even worse, a patronizing view of the farmer or fisherman, and assume they know his or her business better than the operator. Planners and advisers have first to listen to producers so that they can help to resolve problems as producers see them. Even when planners recognize this need it is often difficult for them to give up the old ideas completely. There is the added barrier that, even if individual officials understand the advantages of close liaison with producers, their senior officers in central ministries, departments, and agencies may not.

Regional planning coordination

A third problem relates to coordination of the various regional plans. There is always the danger that competing regional claims may get out of hand and exceed available resources, or result in wasteful duplication of projects in different regions. The finance provided by the central government is a powerful device for coordinating regional and national development plans' objectives, but regional pressures for funds are often so great as to upset the national strategy. Making sure that objectives do not conflict does not necessarily require that either producers or government abandon their objectives. It may, instead, involve compromise about priorities or timing.

Regional planning organization

The need to establish suitable systems for regional planning raises the question of what kinds of institutions are needed to prepare and implement regional and local plans. In most developing countries there are few subnational planning entities. This means that in many of them suitable units must be devised and established if effective planning is to take place.

Whether regional plans are prepared by national or regional bodies generally depends on the stage of development of the regions concerned, the availability of adequately trained and competent regional planners, and the size of the country. As a general rule the earlier the stage of regional development, the greater the likelihood that regional plans are prepared by the central planning agency, or by an authority established for this purpose by the central government. Conversely the more advanced the region, the greater the availability of competent regional planners; and the larger the country the more likely it is that plans are prepared by the regions themselves.

Because regional planning in most mixed-economy countries has been for economically backward areas, with special problems and shortages of financial and human resources, most regional planning has been conducted by central planning agencies, or by special authorities created by national governments. Frequently regional plans have been prepared outside the regions, usually in the state capital, with little participation of the people in the regions. Aside from the fact that this approach has not been very successful, it is less likely to be used today when political awareness in the regions is increasing.

The institutional problem is one of finding the right structure and balance between centralized and decentralized authority, self-financing, and decision-making. Unless properly structured institutions are set up with the right balance, there is a danger that the most logically consistent regional plans will not work in practice.

Lack of regional information

Planning at the national level has been seriously hampered by the scarcity of reliable statistical and other information. Indeed, the lack of adequate data on such basic series as the size and growth rate of national populations, production, investment, consumption, and employment has frequently cast doubt on supply and demand projections, as well as parameters included in national planning models. The fact that regional and local statistics in most poor countries are generally much worse, where they exist, than national data, makes it particularly difficult to plan effectively at regional and local levels. The lack of good data on land use, water resources, and manpower constitutes an especially great handicap. Even in countries where political subdivisions are defined as agricultural regions, as in India, serious deficiencies in data impede effective planning; and the situation is usually worse when an agricultural region or sub-region does not conform to administrative boundaries.

The data gap is thus a fifth problem. Its importance, and the time needed to resolve it, makes it probable that regional and local planning will have to proceed at a lower level of technical sophistication than at national levels, and that, for the time being, mathematical formulations have little practical value. Moreover the lower the jurisdiction is placed in the hierarchy, the more modest are the requirements for comprehensive planning. Nevertheless, as regional planning in such different countries as China and Tanzania has shown, there is no reason for postponing the start of regional planning until adequate data become available. Some specialists become so preoccupied with the data problems that they all but counsel postponement of regional planning until adequate data, in their opinion, are produced.

In the long term, it is necessary to set up suitable systems for the collection of data which adopt uniform units of measurement to allow intra- and inter-regional combination and comparison of data on a continuing basis, and for clearly defined regional and local areas. Therefore, consideration should be given to the preparation of a multi-annual statistical collection plan, including a system of financing for its implementation, which sets forth the types of data to be collected and the priorities to be given to each. As the task of carrying out such a plan is likely to be beyond the capacity of existing central statistical organizations in most developing countries, attention must be given to the reorganization and expansion of the central statistical office, as well as to the possibility of dividing the task between it and regional statistical organizations.

Training regional planners

As regional planning is still in its infancy, regional skills and expertise are in short supply. Because of the variety of skills demanded, training for regional development must be problem-oriented. This implies a multi-disciplinary approach. While much has been written about the need for a 'comprehensive approach' to development problems, little as yet has found its way into training materials and courses. The same is true about planning 'from the bottom up'.

Another training constraint has to do with the 'labour-intensive' way in which planning, as a discipline, has been taught. Teachers in the field are generally highly trained specialists whose teaching reaches only a few students each year. Moreover, would-be planners are generally required to undertake a lengthy course of study to provide them with the ability to formulate sophisticated plans, often based on mathematical models. Training of planners has thus proceeded so slowly that there is still a great shortage at the national level and, even more so, at regional and local levels.

If regional planning is to expand more quickly, a more 'capital-intensive' approach is needed. If, for example, the information which planning specialists have accumulated were to be put within a 'programmed' form, teachers could use these materials to improve their effectiveness.

Typical contents of a regional planning manual include:

- Objectives for regional, sub-regional and district planning.
- National policies and measures which have been or will be adopted to achieve these objectives.
- The emphasis to be given to the different ecological regions by the central government, if the emphasis will differ among them.
- A list of priorities for regional planning from the national viewpoint, with an explanation of the rationale for these priorities.
- Procedures to follow at various levels in preparing plans, programmes, and projects, data to include, and how to present these.
- The cost-benefit or appraisal methodology to use.
- A requirement that plans be prepared with at least two alternative assumptions about their size, one with a higher cost than the other (as specified by the central government), each with a list of projects and programmes to which priorities have been assigned.
- An outline for a regional stocktaking and diagnostic survey.

Figure 3.5 The regional planning manual.

The need for a regional planning manual

The central government usually establishes basic rules applying to regional planning, which all regions should follow. However, governments sometimes have very complicated procedures and requirements, insist on too much detailed information, or do not define with sufficient clarity the technical criteria which regional planners must follow in preparing plans, programmes, and/or other projects. The result is that the whole process is uninformed and disorganized. In order to provide regional, sub-regional, and district planners with an understanding of what needs to be done, and to have them follow the same basic rules, a central government should prepare and disseminate a basic planning manual. The suggested contents of the manual are described in Figure 3.5.

Chapter 4
Preparing an Aquaculture Sector Plan

Synopsis

The chapter describes the preparation of a plan for development. First it summarizes the twelve components of planning, and the sequence in which information is assembled and integrated. The next section describes the stocktaking and diagnostic surveys required, followed by a section detailing the information required for each component of the plan. The final section concerns the formulation of strategies for development, which set out policies, projects, programmes, and other measures to be undertaken, and thus the construction of the development plan itself.

4.1 The components of an aquaculture plan

In the sparse literature on aquaculture planning, there is relatively little about how to devise appropriate strategies and policies for putting a plan into action, how to prepare a stock of suitable projects in sufficient number to enable intelligent choices to be made about them, how to encourage fishfarmers to increase their investment and output, or how precisely administrative procedures must be changed and adjustments made to organizations and institutions – if sector goals are to be achieved.

There is, of course, a good reason for this state of affairs. It is easier to formulate a plan than to state, in sufficient detail:

- *what* action is specifically needed to reach planning goals,
- *how* things are to be done,
- *who* is to be responsible for what,
- *when* (with sequence and latest dates) action must take place if the plan is to be implemented within the planning period.

This does not mean that plan formulation is easy or less important than implementation. It means that insufficient attention is usually given to implementation. As a result, many plans have little chance of being carried out in practice. The two tasks are inseparable. If this was better understood,

- Development objectives.
- A stocktaking and diagnostic survey.
- Demand and supply projections.
- Setting targets and allocating resources.
- Choice of strategies.
- Choice of policy, and its instruments, including projects.
- Research and studies to obtain technical information needed (this is likely to be 'ongoing' work), and extension.
- Consultancy services required.
- A training and education programme for men and women to fill actual and potential gaps in managerial and technical positions (ongoing).
- A programme of public expenditure for financing each year of the plan period, including the sources of finance.
- Changes and improvements needed in organizations, institutions, and administration.
- A system of monitoring, reporting, and control.

Figure 4.1 The twelve components of an aquaculture sector plan.

plan formulation would be even more difficult than it is now – but, at least, it would be more realistic.

A good plan should have many more of its components concerned with implementation than formulation. Figure 4.1 summarizes the 12 major components of an aquaculture plan: four deal with formulation, eight with implementation. These components are reviewed briefly later in section 4.4, and some are dealt with in more detail elsewhere in the book.

4.2 The planning sequence

The sequence in which the 12 components (Figure 4.1) are stated has significance. Some things should be done before others, even if only tentatively. The setting of the sector's development objectives, for example, should precede the stocktaking and diagnostic survey because a given objective may, and often does, require specific information.

Setting the sector's objectives should also precede target fixing (as targets are really quantified objectives). Similarly, the implications of new projects, and programmes for organization and administration, should be thought out before organizational responsibilities are assigned, and requirements for technical assistance determined.

Unfortunately a rational sequence is not always followed. In practice sector surveys are often made and targets set before objectives are clearly defined. Economic policies and measures are adopted without reference to the requirements for achieving targets, and new agencies established, or existing ones reorganized, before the tasks which must be carried out to implement the sector programmes have been properly identified.

Although some aspects of planning precede others, it would be a mistake to suppose that no deviation is permitted from the suggested sequence, and that each component must be dealt with consecutively in the order

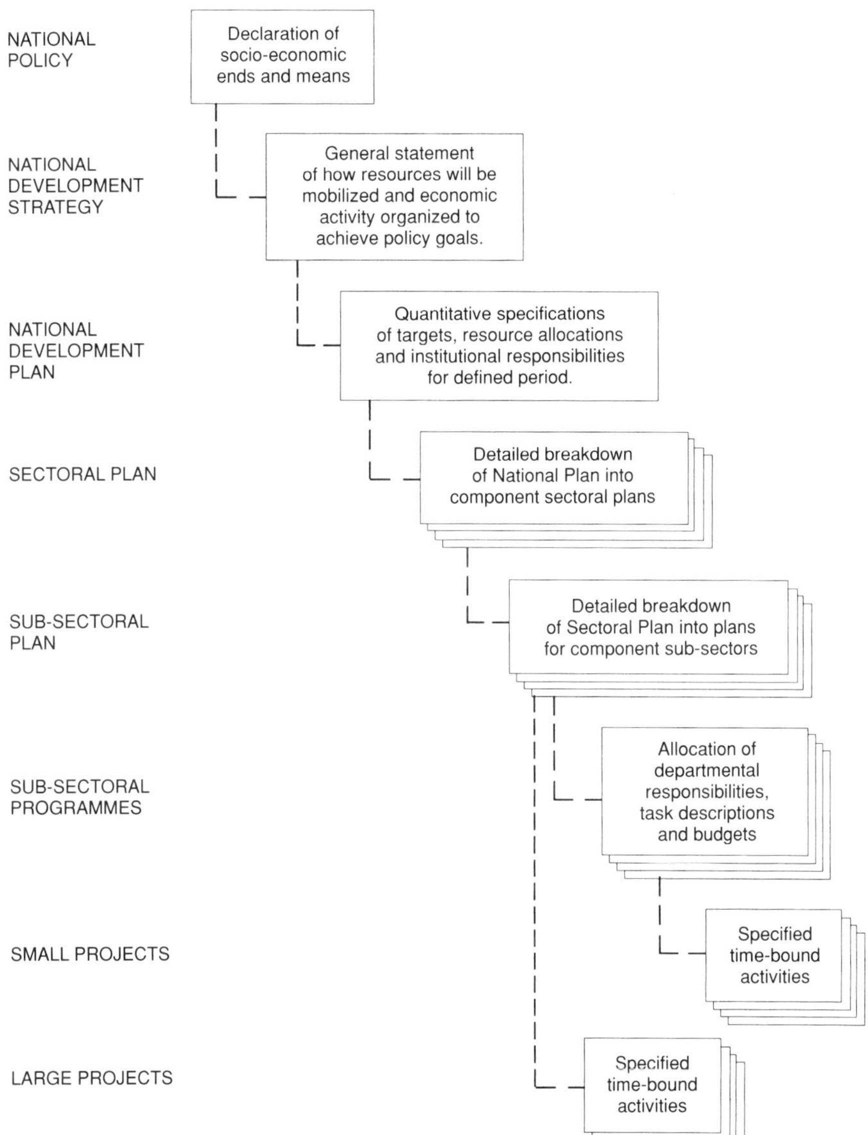

Source: FAO, 1986.

Figure 4.2 Schematic representation of the planning hierarchy (Source: FAO, 1986).

given, one at a time (see Figure 4.2). Work on some components low in the list may have to begin before others, higher up. Such is the nature of planning. Project identification and preparation, for example, may take two or three times as long as the preparation of the entire aquaculture plan.

Usually, work must proceed simultaneously in two or more components, with the emphasis changing from one to another. This is because sector-like macro-economic planning is a conscious process, with interrelated and mutually interacting elements. Each component must be coordinated with the others if the plan is to be consistent. Targets may have to be altered as planning proceeds if, for example, strategies or policies required to achieve them are politically not feasible. Projects may have to be reduced in number and size.

Strategy may have to be modified in the light of rates of return realizable from available projects, and even objectives revised or discarded if, for example, investment funds prove to be insufficient or individual commodity prices undergo a sharp change. Conversely, additional funds may have to be found to permit retention of objectives and projects considered of the utmost priority.

In short, planning for aquaculture starts with assumptions and speculations which may turn out to be untenable, and by iteration, or successive approximation, moves towards what is possible. The final form of the plan emerges at the end as one would expect.

4.3 Integrating the planning components

Establishing the objectives of the plan

Objectives chosen for aquaculture development must be considered provisional, subject to change. Planners have the responsibility of providing those who decide objectives with guidelines to help them make rational choices. Information given to the decision-makers, whoever they may be, should include a wide range of data about the main aquaculture problems of the country and its regions, feasible alternative solutions, and estimates of resources needed and likely to be available for this purpose. Such problems may include high rural unemployment or underemployment, a greatly skewed pattern of income distribution among farmers, fishfarmers and rural people, low levels of output and productivity, lack of foreign exchange due to poor levels of income from agricultural and aquacultural exports, etc. Information about such problems is usually adequate in most countries for preliminary objectives to be set.

Planners must realize that economic objectives must conform to social and political realities and aims, rather than the other way around. In aquaculture in particular, certain governments have tried to limit development opportunities to local groups.

Economic objectives may have to conform to social objectives and political realities, even at the expense of economic efficiency

In shrimp aquaculture in Mexico, and extensive lagoon aquaculture in Greece, the respective governments have attempted to direct the benefits to local populations by decreeing in law that only cooperatives can participate in these forms of aquaculture production.
Norway attempts the same by limiting the legal size of farms and allocating licences to areas of the country where the need for economic opportunity is perceived to be greatest.

Planners frequently find it difficult to accept this situation as it may lead to higher economic costs, or lower returns, than if economic costs alone were paramount. This attitude misses the point since the main purpose of planning is not to maximize cost-effectiveness but to achieve social and political objectives at the lowest possible cost.

Just as objectives cannot be left to national planners to decide, they cannot be left to foreigners either. This is because the way in which development, and hence its objectives, is defined depends, at least partly, on a country's values, culture, and heritage.

Objectives and programmes based on foreign cultural values may generate change which is incompatible with traditional institutions and patterns of behaviour. In that event the changes are likely to have only a temporary effect (while the foreigner is around) or, if more permanent, may have an unsettling and harmful influence on the traditional society.

It is not always tradition or the 'dead hand of the past' which shapes a people's attitude. The experience of a people may play as powerful a role as tradition in shaping social, political and economic attitudes. Every society should therefore set its own development objectives, including, of course, those for aquaculture. Foreign advisers may have a useful and supporting role to play in this process.

Recognizing the importance of a few clear objectives

It is no simple matter to get an unambiguous statement about objectives. They should preferably be expressed in concrete rather than abstract terms because vagueness leads to confusion. Political leaders are sometimes reluctant to commit themselves to clearly stated objectives which may be seen as a threat to powerful interests. They are also likely to wish to satisfy every important group by fudging the issues, in which case planning objectives becomes an exercise in public relations rather than the real thing.

It is common, therefore, to find sector plans with an abundance of objectives, some of which quite plainly are contradictory. A long list of objectives is generally harder to achieve than a short one because of the scarcity of resources. Priority becomes harder to determine. As the problem is more often one of too many objectives than too few, one precept worth bearing in mind is to have as few objectives as possible. In aquaculture plans it is normal for objectives to be stated in a confusing way. Statements of objectives are often political rather than realistic economic and/or social objectives.

Aquaculture objectives are often political statements rather than realistic economic and/or social objectives

Objectives of the State of Hawaii's aquaculture development plan include:

- Provision of more, and a diverse range of, employment opportunities.
- Diversification of the economies of all Islands.
- Greater self-reliance in food supply by substituting locally farmed aquatic products.
- Improvement in the balance of trade by increasing exports of aquatic products.
- Reduction of development pressure on prime agricultural lands.
- Assurance of opportunities for rural residents to preserve their life-styles.
- Maintenance of open spaces.
- Enhancement of the State's quality of life, involving economic, social, cultural, and environmental components, by the wise use of Hawaii's natural and human resources.

Source: Department of Planning and Economic Development, State of Hawaii (USA), 1978.

Distinguishing between 'ends' and 'means'

The relationship between ends and means is illustrated in Figure 4.3. A related principle is that ends, or objectives, should be clearly distinguished from the means to achieve them. There is often confusion on this point. For example, where self-sufficiency in fish supplies is the objective, a programme to improve fish quality may be the strategy; the use of government subsidies for the improved stock may be the policy; and the sale of this improved seed by cooperatives at subsidized prices may be the policy instrument. The quality improvement, the use of subsidies, and the role of cooperatives to sell the subsidized seed should not be included as objectives because they are the means for achieving self-sufficiency. Such confusion is not hard to find in attempts to plan aquaculture.

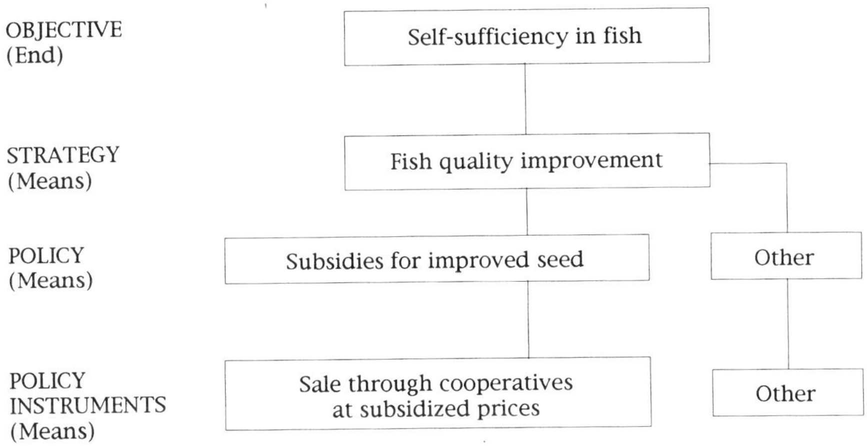

Figure 4.3 An illustration of 'ends' and 'means'.

Objectives and the means of achieving them are often confused in planning statements

'To date, all the Provinces and the Territories, as well as the Federal Government, have had various measures to encourage commercial aquaculture development in Canada. Future development will be achieved through the combined efforts of industry and government working towards common GOALS. These GOALS embrace the following:

- To encourage the development of commercial aquaculture in Canada in a manner that is complementary to the continuing development of the wild fishery.
- To increase the economic returns from intensified production and harvest of high value, marketable species of fin fish, shell fish, and marine plants in the regions of Canada.
- To improve the quality and expand the variety of Canadian fish products.
- To improve the reliability and of the supply of Canadian products to be marketed in Canada and abroad.
- To create new employment and enriched income opportunities in the production of fish.
- To encourage long-range stability in the country's fish production sector through diversity and continuity of supply.
- To promote the development and application of the most advanced technologies for intensive production and marketing of fin fish, shell fish, and marine plants across Canada.'

Source: Science Council of Canada, 1984.

Identifying distinctions between the objectives

Social and political objectives are becoming a regular feature of development programmes. As objectives, such as raising income levels in depressed rural areas, may reduce the rate at which output may otherwise increase, in the short run at least, it is essential for the political leadership to indicate the extent to which it is willing to sacrifice growth of output and income to achieve equity objectives. Otherwise, it is difficult to select appropriate strategies and policies.

Short, medium, and long-term objectives should be clearly distinguished. A realistic assessment of what can be done in the short or medium term can usually be based on recent performance, available resources, and the current level of administrative competence. The rest is best left for the more distant future, broadly framed, and indicating only the general directions in which to move.

Structural change falls into the latter category because it tends to take many years to achieve. Such change involves items such as the shift to modern commercial aquaculture, greater diversification of output, a large reduction in the proportion of the population dependent on agriculture (including aquaculture), and a shift in dietary habits.

4.4 The twelve components of aquaculture planning

Stocktaking and diagnostic surveys

A stocktaking and diagnostic survey is needed early in the planning process to provide information about the wide range of factors influencing sector performance (section 4.5 of this chapter outlines the information needed). The scope of this survey should be limited and closely allied to plan objectives. Good planning can be done with surprisingly few data, provided they are accurate, timely, and relevant. Conversely, large amounts of inaccurate, untimely, or only partially relevant information can be costly and of little help to planners. Great care is therefore needed in setting up the survey. The survey team may well include foreign advisers, whose role should be not only to assist in carrying out the survey to a high professional standard, but also in training nationals. The latter may then provide the nucleus of a planning unit.

Planning requires knowledge about recent progress in aquaculture, the existing situation (especially the main problems impeding its development), and its potential. All this information is needed for re-assessing current investment and other development activities, as well as for planning new measures, and setting benchmarks against which to monitor progress. Improvement in the way things are being carried out can often make as great a contribution to development as the start of new activities. Analysis of recent and current conditions may also reveal structural and institutional relationships, or policies, which have to be modified.

Knowledge of the recent past, and the present, is thus essential for defining the nature of the planning task and its limits. A stocktaking and diagnostic survey, carried out early in the planning process, is a way of providing this information.

This particular survey is quite different from a basic research study, although short-term research and a few fact-finding *ad hoc* studies (which can be completed quickly) are not necessarily to be excluded. A stocktaking and diagnostic survey should be, as its name implies, a broad-gauge inventory, and an assessment of largely existing information about the sector or subsector.

Information is never available in sufficient quantity and quality to meet all planning requirements, and those responsible for the survey must resist the temptation to produce new data which take too much time to obtain. Instead they should concentrate on the collection, analysis, and projection of *available* data. At the same time, they must note the information gaps which require longer-term research and study, so that provision may be made to fill these gaps with a programme of pre-investment and other research studies. This may require years to complete.

The data collected, the way they are arranged and analysed, and the choice of issues examined must be dictated by the plan's objectives. The survey is not intended to be a means for accumulating and analysing all available facts about the aquaculture sector. What matters is not the quantity of data but their relevance. Careful discrimination is needed in selection and rejection of data to be acquired.

The scope of the survey should therefore be limited, and the survey report focused around particular issues rather than discursive. The information in the report must be pointed and relevant, not merely of general interest, and sufficiently detailed to be useful. It should not be so detailed that the report is too bulky for easy use as a reference.

A good way to begin is to outline the scope and emphasis of the survey, as well as the sub-sectors to be covered. As far as possible, it should provide separate information for each of the major regions in the country. An early task is the collection of all previous studies (and plans) of the aquaculture sector, as well as all available indexes of aerial (and satellite) photographs, and topographic and other relevant maps for examination, comparison, and analysis.

In many countries there are surprisingly good opportunities to obtain regional or local plans, statistics, maps, and other useful material from district and other area offices. Those conducting the survey should therefore arrange to visit these offices. It is also desirable, as soon as possible, to determine the subjects for which short-term field investigations will be required. This is not always easy to do, unless gaps in information are known in advance, or become clear soon after the survey begins. But the later such studies start, the smaller the chances that they will be completed in time for use in preparation of the programme. In this regard, surveyors should always avail themselves of the vast published resources of FAO. These include technical papers published by the headquarters in Rome, and the many projects executed by FAO all over the world. For example, the work of Tacon *et al.* (1987) documented feed resources in the countries of Latin America and the Caribbean, providing information indispensable to planning sector projections for growth. FAO documents are available at its library in Rome, and in the offices of the FAO representatives in member countries.

Finally, it is important to set up a timetable, with deadlines, for each phase of the survey and to check progress at regular intervals. By 'planning the planning' this way, using some form of critical path analysis (as proposed in *Aquaculture Project Formulation,* by Insull and Nash, 1990) it is easier than it would otherwise be to take appropriate action if work falls behind schedule.

Demand and supply projections

Demand and supply projections, however crude or sophisticated, are an important feature of aquaculture planning (see the examples opposite). Yet currently there is a lack of well-prepared projections of this kind. Using different scenarios, they can indicate the extent of the effort required from domestic aquaculture to help satisfy demand from home and abroad.

Projections of demand are required for food fish and feed materials and for domestic and foreign markets. The main determinants of demand are population growth and changes in *per caput* income. Demand for food fish tends to expand in line with population growth. As incomes rise expenditure rises too but not proportionately, and this income response differs for individual foods. Estimating domestic demand for different kinds and qualities of fish thus tends to be difficult. A compromise has to be found between the various techniques of estimation and available data.

It is in rural areas where nutritional improvement in diets is of pressing concern. Demand from the rural poor for the products of aquaculture is likely to increase where the culture of fish is known and appears as a preferable substitute in the diet. As 'industrial' aquaculture has not yet produced fish affordable by most of the urban and rural poor, it seems unrealistic at this time to hope that industrial types of aquaculture will be able to solve nutritional problems on a large scale in developing countries.

There are likely to be special problems when estimating demand for raw materials, as markets for their various end-users have to be assessed separately. And similarly for export demand, where estimation techniques are likely to vary according to the nature of the commodity and the provisions under which it is exported.

Projections of output and supply are technically much more difficult to make than those of demand. There is the uncertainty of fishfarmers'

Understanding market influences and using forecasting information on supply and demand projections assist planning

Extracts from three background planning documents state:

(a) 'Are Atlantic and Pacific salmon substitutes? With the increasing demand for fresh salmon, especially in winter months, an unresolved empirical issuc is whether farmed salmon (Atlantic and Pacific) are strong substitutes for wild Pacific salmon. Alternatively, do fresh farmed salmon and fresh wild Pacific salmon compete in the same markets? The question can be further complicated by asking whether fresh pen-raised Atlantic salmon (from either domestic or foreign producers) compete with pen-raised Pacific salmon (chinook or coho) and which may also compete with wild Pacific salmon (chinook and coho) from the capture fisheries? And finally, can fresh farmed salmon produced in the U.S.A. compete with fresh farmed salmon produced abroad and sold in the U.S.A.?'
Source: US Department of Commerce, 1988.

(b) 'Future production. Most Member countries report that production will increase and that further support in the form of subsidies and research will be provided. It remains, however, that most aquaculture production will continue as private business enterprises and as such factors to be looked upon in forecasting further developments may be somewhat different than just a simple extrapolation. Any forecasting should include an assessment of factors such as:

- suitable sites with due regard to possibilities for licensing,
- sufficient supplies from support industries and two in particular, namely (a) hatcheries, and (b) fish meal,
- cost developments, such as feed components, and investments,
- market developments,
- resources available from the public to support investment and research.'

Source: OECD, 1989a.

(c) *Table*: Estimate total market potential for farmed salmon in 1990 (all figures in tonnes).

Region	Low	Probable	High
France	21 000	28 000	35 000
Germany	16 500	22 000	27 500
UK	11 250	15 000	18 750
Rest of Europe	11 250	15 000	18 750
USA	22 500	30 000	37 500
Japan	7 500	10 000	12 500

Source: OECD, 1989b.

response to changing prices, costs, and technology, added to such factors as weather and diseases. Then, output depends so often on decisions of many thousands of small and medium-sized producers. The impact of various legislation on fishing practices has also to be considered, not always an easy calculation to make. Nevertheless, experience has shown that it is possible for planners to provide useful estimates of future output or catches, under given assumptions of yield, of fishery resources, input of feed, and fertilizer, etc. Projecting supply takes account not only of domestic output but also of imports, exports, and any changes in stocks.

Once demand and supply projections have been made it is necessary to bring the two into balance. Some sort of commodity balance sheet may be constructed as a basis for determining what adjustments are needed. They may require a considerable change in the proposed aquaculture programme, or possibly only minor alternations. In any event, an element of flexibility should be a feature of the plan eventually adopted.

Setting targets and allocating resources

Clearly defined objectives are a feature of good planning. When these are quantified in specific terms, backed up by policies, investments and other proposed actions, they become targets. As social and political factors play a crucial role in target selection, politicians have the final say in this matter. Planners have the task of explaining what is involved in choosing the various targets.

Setting targets for aquaculture plans

'Target 2000 Status. It is necessary to establish some target towards which the efforts of the Plan would be directed. There is no clear definition of the carrying capacity of the (Cook) Inlet. Additionally, to all but the managers, the most meaningful number is the one that describes the harvest goals for the year 2000. After considerable review of historic and current trends and levels of harvest by all user groups a target of 12,000 million salmon was adopted. This mark, which is about 50% higher than the best total harvest ever recorded in the Inlet, is both high enough to necessitate a more thorough understanding of the salmon and of the Inlet, and modest enough to be within reach, if all identified projects proved both feasible and successful.'

Source: Alaska Department of Fish and Game (USA), 1982.

Target setting is a difficult art. Set too high, targets may become very discouraging or even meaningless, and too low a waste of resources and output. Targets are sometimes confused with projections, but they are quite different. Care is needed in deciding upon the number of targets. Generally, the fewer the better, and they should be flexible enough to take account of changed circumstances. Moving from projections to targets involves a process of iteration, or successive approximation, in an attempt to harmonize demand, supply, and resources. Several methods are available to deal with this problem. Basically they all involve some attempt, however feeble, to allocate national resources to best advantage. It is not an exact calculation, because reliable data on social costs and returns in the agricultural (and aquacultural) sector particularly, are generally inadequate. Even so, and even in the poorest of countries, this effort should be worthwhile.

Targets should always try to take account of regional and local differences, although excessive disaggregation can be self-defeating. Target setting is a two-way process, with planning from bottom upward and from top downward. But sector input and output targets need not be the totals of all regions or localities. National objectives have priority, and some adjustment may be required in local and regional target-setting. Those responsible for achieving production targets should have a voice in setting them.

Choosing the strategy

Policy making and analysis are the heart of planning, but the choice of suitable policies can be a complex matter. A frame of reference is needed, and this is what a development strategy should provide. Often, however, there is a total absence of a policy framework as well as confusion between objectives, strategies, and policies. Different types of strategies may include short- versus long-term emphasis on small enterprises, import substitution, export expansion, and diversification. Various constraints limit the choice of strategies and there is need for a systematic analysis to arrive at a compromise between typically conflicting economic and social aims, and to secure effective use of the scarcest resources.

A vital step is to recognize what the major constraints are. They will normally change as a country develops. There are different constraints at national, sector, local, and enterprise levels at any given time. The best way of dealing with this 'restricting' problem is to choose a strategy or strategies least bound by constraints; take full advantage of relatively abundant resources, and use most carefully the scarcest resource. If development proceeds according to plan, strategies will gradually change to adjust to a new resource situation. Whatever strategy is adopted its full implications should be carefully thought out in relation to expected changes in, for example, outputs, inputs, incomes, and their distribution and other social consequences. This subject is discussed in more detail later (see section 4.6).

Understanding the constraints is necessary if realistic development strategies are to be formulated

Aquaculture Development Constraints in South African Development Coordination Conference (SADCC) Countries of Africa. Being a relatively new husbandry system to SADCC countries, aquaculture faces a number of constraints in its development. The constraints vary from country to country, and opinions on how to eliminate or reduce these constraints also vary. However, the approaches selected to deal with these constraints have a direct bearing on the aquaculture development policy of the country. Therefore it is important to analyse critically the development constraints during the process of policy formulation and planning, a process which most African countries are now involved in. The following is a list of major constraints:

(1) *Environmental:*

- unavailability of suitable land,
- unavailability of good quality water,
- water security and multiple usage of water resources,
- pollution problems,
- environmental degradation.

(2) *Biotechnological:*

- lack or shortage of fish seed supplies,
- unavailability or high costs of fish feed,
- adoption of aquaculture technologies without local trials,
- shortage of trained manpower.

(3) *Financial and economic:*

- underdevelopment of national infrastructure (roads, communications, etc.) which isolate potential production areas from markets,
- public sector's lack of long-term capacity to support aquaculture, especially programmes started by donors,
- inaccessible credit facilities,
- insufficient marketing facilities,
- high opportunity costs of inputs, even in rural areas.

(4) *Social and human:*

- fish consumption habits and patterns; aquaculture usually develops in areas far removed from influence of capture fisheries, and these are areas with low demand for fish due to unfamiliarity and taboos,
- common property and common ownership problems,
- lack of analysis in identifying target groups in traditional

rural set-ups for introduction of rural aquaculture and farmer training.

(5) *Administrative and legal:*

- absence of, or inadequately prepared, government policy,
- lack of, or low, administrative support for aquaculture by Departments of Fisheries or responsible Ministries,
- inappropriate administrative organization in some countries,
- problems of land tenure and access to water resources in inland aquaculture,
- legal problems on user rights and access to coastal waters for marine aquaculture.

Source: FAO, 1988a.

Setting policy and selecting policy instruments

Key policy decisions involve what is to be attempted and how objectives are to be achieved. In any given country policy makers will come up against constraints. It is their job to identify and, if possible, overcome them. There is likely to be a range of options open to them, from direct measures to increase sector output and productivity, to less direct options such as price and tax incentives, institutional changes, credit, and marketing improvements, etc. They will also have to develop policies which tackle very awkward interrelated problems of unemployment and underemployment in rural areas, mechanization, and the use of appropriate technology.

The various types of measures used in sector development make different demands on resources. The most scarce are likely to be foreign currency and skilled personnel, particularly administrators and managers. Care is therefore needed to put together the right bundle of measures to take full account of both abundant (often unskilled labour) and scarce resources. Time needed for the various measures to show results must be weighed carefully. Too often, slow-maturing projects and measures take the lion's share of funds at the expense of current, and possible future, improvements. A time-sequence of policies needs to be established early, together with priorities for the different measures, as part of the continuous process of policy planning. The common tendency to choose complex all-embracing solutions must generally be resisted. Decision-making can be assisted, by pilot projects, trial programmes, tackling only parts of a problem, and decentralization. The final question may well concern which comes first and which later, rather than what to leave out.

Projects

Identifying, preparing and successfully implementing aquaculture projects is often a difficult exercise, especially in developing countries. Failure is common, or only partial success may be achieved. It is necessary to examine why good projects are so few, and how this situation can be improved.

At the outset a distinction needs to be made between experimental, pilot, demonstration, and production projects. Then comes the project cycle with its six main parts. First, a project has to be identified, possibly using a special or sector survey, and taking full account of people's expressed and often half-articulated needs. Formulation or preparation comes next, when the main aspects of the project have to be examined – technical, financial, economic, institutional, managerial, environmental, and commercial. Techniques used in this exercise are well-established and reviewed later in Chapter 10. All too often, technical feasibility is taken as the key factor. Managerial and institutional factors are commonly underrated – to the detriment of a balanced assessment. Project appraisal, the next step, remains more an art than a science, regardless of the quality and quantity of the assembled data. Success in implementation, the fourth stage, depends mainly on good management and organization (using the logic frame and critical path analysis), and close alignment between a project's special requirements and the facilities provided, especially at local level. Special arrangements are often needed, too, to manage and supervise large projects either by expansion of existing bureaucracies, creation of autonomous units, an organization which reports to a central, regional, or field agency, or use of an external agency. A critical element in project management is monitoring of results achieved against those planned for – the fifth component of the cycle.

Finally, there is the sixth element – evaluation – whose main aim, both 'ongoing' and 'ex-post', is to enable planning to be improved (see Chapter 9). The ultimate test may be whether or not a project generates its own continuing and expanding activities. Immediate and tangible results may be much less significant than later less obvious consequences.

Research and extension

Aquaculture development is handicapped, particularly in countries with little experience of it, by inadequate support for national research, often misdirected.

Production of better and more relevant research requires close cooperation between planners, extension and research workers, and fish farmers. Priority should be given to the real problems of the farmers, which is not always the case. If it were, more emphasis would be given to the holistic approach to fishing enterprises – looking at the whole enterprise, including its socio-economic parameters as well as the usual emphasis on its technocratic elements. Research on better management of suspended solids and increasing efficiency of feeds, for example, is likely to reduce the incidence of pollution caused by aquaculture; and developments in fish nutrition and improved knowledge of care and prevention of fish diseases are likely to be very helpful, especially in hatchery management. Research on genetic engineering, on the other hand, is not expected to be a major contributor to aquaculture, compared with the very positive effect of high yielding grain varieties (HYVs) in terrestrial farming.

A key role in aquaculture research rests with organization and the coordination of all the interested parties. Research in most developing countries is mainly done by government departments. As indicated above, however, programmes tend to be biotechnically oriented and concerned primarily with hatchery production of selected species under different levels of husbandry. Such research faces several constraints, such as inadequate funding, inexperienced research staff, weak management, and the predominance of biotechnicians. This leads to major constraints with little successful fundamental research being carried out on some species of importance (such as milkfish, grouper, and mullets) and on species of local importance in tropical regions; uncoordinated work in countries with little aquaculture (limited to little more than reporting of observed events, rather than scientific enquiry to establish fundamental cause-and-effect relationships); and with the dominance of biotechnology there is a tendency to overlook the interests of producers and consumers. Technically-adapted culture systems are promoted with insufficient attention given to the system's social and economic consequences.

Extension workers can play a significant role in acting as a channel between research and production, a two-way channel. In most countries (with the exception of some with well-established aquaculture sectors, such as Hungary, Israel, Japan, and Thailand), this service is still weak, with governments underestimating the importance of appropriate and adequate equipment, especially transport. There also seems little concern about which methods of extension are most effective and efficient; for example, on-farm demonstrations, the training and visit (T&V) system, or the use of 'model farmers'.

Poor funding is often the problem for extension, leading to inadequate and (often) badly-trained workers, heavy workloads only partly caused by their educational duties. Advice is mostly technical (dealing with water management, pond construction, appropriate feeding ratios, maintenance of broodstock, etc.) but often not placed in a proper economic context. Frequently, advice on general management (including marketing, post-harvest handling, and accounting) is not provided or encouraged. Weak extension services are a major handicap in the development of aquaculture, yet there is plenty of evidence from countries well provided with extension that it can give high rates of return (in combination, of course, with an effective and efficient research programme).

The need for consultancy services

Developing countries have for a variety of reasons found it useful to have the services of a wide range of consultants, or technical experts, and this demand is not likely to disappear for some time. They provide specialized skills and knowledge, are generally available for flexible time periods, can give objective assessments, and provide a basis for transferring their skills to national counterparts. They provide no guarantee of improved results. Much depends on a consultant's terms of reference and the phrasing of their instructions.

In the aquaculture sector consultants work in four main areas: policy studies; plan, project, and programme analysis; management; and technical improvement of operating units. Their usefulness in each of these field depends to a great extent on close cooperation with and support from their clients.

Education and training

People with a broad knowledge of aquaculture are needed in government and private agencies to promote realistic aquaculture policies and operational plans. Relatively few are needed but their role may be crucial. This was recognized categorically in a recent comprehensive Thematic Evaluation of Aquaculture by UNDP, FAO, and the government of Norway.

Each country needs people educated with a broad knowledge of aquaculture

'Ideally, some government (and private agency) staff should be given the opportunity to obtain an education in aquaculture as the beginning of any effort to develop a strategy for its development. There is a diversity of interest among governments concerning education and training, and the emphasis placed on each. While many people in Latin America and Asia have received an aquaculture education, few have had similar opportunities in Africa. There would seem to be little need for further general aquaculture education in Latin America until more is known about the outlook for the sector in that continent. On the other hand, there remains a shortage of trained specialists (natural and social scientists, engineers) in subject areas supporting aquaculture.'

Source: FAO, 1987.

Opportunities for academic education in aquaculture are few, and mostly found in industrialized countries and a few places in Asia and Latin America. Training in aquaculture skills, however, is a prime function of government departments (unless countries wish to maintain dependence on external advisers).

A developing country without a well-established aquaculture sector faces two main difficulties when it attempts to train aquaculture technicians and extension workers: a shortage of qualified teachers, and doubts about the validity of permanent training institutions in view of the relatively small number of trainees initially required. Regional cooperation can help in this regard. Also, the need for special institutions may be minimized by using fish culture stations as training centres and, where regional cooperation is not feasible, expatriate teachers can be used.

Financing the plan

Financing the aquaculture plan, including a programme of public expenditure for the financing of each year of the plan period, can be a major problem. Although foreign assistance can and usually does provide an important contribution to the total funds needed, the bulk comes from domestic sources. There is thus the important problem of coordinating the flow of funds from both sources so that development is not held up, for example, by late or non-arrival of funds, or poor scheduling of activities. This points to the need for extreme care in the costing of projects and programmes (for example, see Figure 4.4, details of the Aquaculture Development Plan for the State of Hawaii, USA) and the ability to make adjustments to them if financial problems arise, as they often do. Attention has also to be given to the means of raising domestic funds for aquaculture development.

Mobilizing rural savings and creating new and better rural credit institutions are not easy tasks because of the very nature of agricultural (including aquacultural) systems and rural power structures. The decentralization of fiscal authority can be a helpful step in this direction. Other positive developments may include growth of private rural banks and local cooperatives, with emphasis on the need to mobilize savings (however small) and improve the links between formal and informal financial markets.

It should be remembered that one test of a plan is whether adequate finance is available to put it into action. If not, the plan becomes little more than an empty political gesture.

Organization and administration

Weak systems of public (and private) administration hinder the preparation and implementation of economic and especially aquacultural development plans. Traditional structures able to cope with mainly routine official business are unlikely to satisfy the demanding criteria of organizations for development. There are several ways of dealing with this problem. The least satisfactory is likely to be wholesale reform of government administra-

Activity		FY 1978–79		FY 1979–80		FY 1980–81		FY 1981–82		FY 1982–83	
		OP	CIP	OP	CIP	OP	CIP	OP	CIP	OP	CIP
Management	C	50	—	93	100[a]	89	1962[b]	90	—	95	—
	N	—	—	—	100	50	1962	55	—	60	—
Planning	C	34	—	41	—	42	—	45	—	50	—
	N	—	—	41	—	42	—	45	—	50	—
Species projects (performance evaluation and development)	C	513	—	576	—	548	—	550	—	577	—
	N	688	—	750	—	1000	—	1100	—	1154	—
Loans[c]	C	—	—	500[d]	—	—	—	—	—	500[d]	—
	N	—	—	—[e]	—	—[e]	—	—[e]	—	—[e]	—
Extension/advisory information	C	146	—	173	—	167	—	170	—	180	—
	N	50	—	75	—	234	—	340	—	360	—
Analytical services	C	40	90[f]	92	—	89	—	90	—	95	—
	N	—	—	184	—	178	—	180	—	190	—
Market development and reporting	C	35	—	58	—	63	—	65	—	70	—
	N	—	—	50	—	126	—	130	—	140	—

Hatchery facilities	C	72	330[g]	81	—	79	—	80	—	84	—
	N	40	—	60	—	158	—	160	—	168	—
Outdoor research/demonstration facilities	C	—	1250[h]	—	—	50	—	100	—	105	—
	N	—	—	—	—	50	—	100	—	105	—
Laboratory research facilities	C	30	—	46	—	47	—	50	—	54	—
	N	20	—	30	—	47	—	50	—	54	—
Total	C	920	1670	1660	100	1174	1962	1240	—	1810	—
	N	798	—	1190	100	1885	1962	2160	—	2281	—

Notes: Operating (OP) and Capital Improvement Project (CIP) funds in thousands of dollars with 'C' denoting State funds and 'N' denoting Federal and other non-State funds.

[a] Plans and designs for the construction of facilities at the AFRC, Sand Island, Oahu, to provide facilities for administrative, planning, species projects, extension/advisory, analytical, market development and reporting, and laboratory research sevices, equipment, and personnel.

[b] Construction of the facilities in 'a' above and purchase of the necessary equipment.

[c] Remains in the Farm Loan Division, Department of Agriculture.

[d] Additional appropriations to the Aquaculture Loan Fund.

[e] Federal loan program is expected to be established.

[f] Remodeling of the existing AFRC laboratory and purchase of water quality equipment from the AFRC. Project began in FY 1977–78 and is expected to be completed in FY 1978–79.

[g] Enclosure and thermo-regulation of the existing AFRC prawn hatchery. Project began in FY 1977–78 and is expected to be completed in FY 1978–79.

[h] Plans, site selection, site acquisition, designs, and construction of the Tropical Aquaculture Center and purchase of the necessary equipment.

Figure 4.4 Five-year development plan (activities, budget, and timetable); DPED, Hawaii (1978).

tion. Instead, it could be improved, gradually changed and integrated with new organizations. This institution-building needs careful watching to avoid unnecessary conflicts of interest and the worst excesses of bureaucracy. Its success is likely to depend on timing – not too hasty or too slow – but taking advantage of gradually increasing numbers of better trained administrators, nearly always in scarce supply. Even in countries such as the USA and Canada there are problems of effective organization of their aquacultural industries primarily arising from socio-political conflict and legal uncertainties rather than the lack of trained managers and administrators.

The ministry of agriculture or of fisheries, as the case may be, is likely to play a key role in development planning, and its form and functions will need to change with the times. An Aquacultural Coordinating Council – of some type – is necessary to bring together the efforts of the various agencies involved in aquacultural development, assisted by programming units within these agencies. While the functions of the ministry of agriculture (or fisheries) will vary considerably from country to country, policy, planning, and coordination are among its main functions. Effectiveness depends to a large extent on strength in the field, with close liaison vital between headquarters, the regions and local areas. Fishfarmer organizations can play a very useful participation role in planning organization. Their effectiveness may largely depend on official efforts at decentralization, and communications at all levels in both public and private sectors.

Effective communications at all levels in both public and private sectors are important

'AGENDA FOR ACTION. Extend assistance to the industry for an initial five-year period to assist in the creation of effective producer organizations to provide a forum for continuing consultations with government, to cooperate and share knowledge within the industry, and to enable producers to develop self-discipline concerning quality assurance standards and market promotion.'

Source: Science Council of Canada, 1985.

Note: The Canadian Government is investing in the development of producer organizations so that beneficial industry–government relations will be established. Such a structure also provides the industry with an internal infrastructure for undertaking its own development.

Aquaculture administration is particularly complex, and coordination between departments is important. The composition of the Joint Subcommittee on Aquaculture in the USA illustrates the range of departmental involvement.

Good communication between government departments is important for aquaculture planning

The Joint Subcommittee on Aquaculture (JSA) is established within the Office of Science and Technology Policy. Members of JSA include:

- The Secretary of the Department of Agriculture,
- The Secretary of the Department of Commerce,
- The Secretary of the Department of the Interior,
- The Secretary of the Department of Health and Human Services,
- The Administrator of the Agency for International Development,
- The Chief of the Corps of Engineers,
- The Secretary of the Department of Energy,
- The Administrator of the Environmental Protection Agency,
- The Governor of the Farm Credit Association,
- The Director of the National Science Foundation,
- The Administrator of the Small Business Administration,
- The Chairman of the Tennessee Valley Authority.

Source: US Department of the Interior, 1983.

Traditionally, the department involved with fisheries gives priority to the capture fisheries sector rather than to aquaculture. This reflects the declared policies of most governments, which make reference to fishing vessels rather than fish farms, and fishermen rather than fishfarmers. Nevertheless, countries with significant or potential aquaculture production do provide such support services as legislation, research and development, training and extension, credit, post-harvest handling, marketing, and production inputs.

Monitoring, reporting and control

Monitoring of planning activity is a useful management system if properly carried out. It continues to get little attention in many aquaculture programmes and projects. Efforts in this direction have often, unfortunately, not been very successful. Too frequently, observation and recording of events have not been followed by appropriate management action. Relatively new techniques are now available to help managers check progress or, otherwise, toward planned objectives; but emphasis tends to be placed too heavily on causes of failure rather than success.

Monitoring requires a reporting system to select, analyse, and present relevant material. This system should meet several criteria, such as generating the minimum information necessary for management to take timely and appropriate action, report both good and bad results, and not require excessive preparation by line and staff managers. Control of plan or project activity based on the reporting system may involve not only traditional management tools, such as budgeting and auditing, but new procedures to take account of a project's special characteristics. Practical difficulties in monitoring commonly arise from poorly designed or inappropriate systems, inability to undertake baseline studies early enough, delays in preparation and analysis of reports, etc.

Costs of monitoring are not insignificant, amounting to as much as 4–5% of total project costs, so an efficient system is needed to justify them. Finally, care is always needed in selecting the projects (or part of each) which are to be monitored.

4.5 Information needed for a sector plan

As noted in the previous section, the 'stocktaking and diagnostic survey' is based largely on the objectives of the sector plan. It is not intended as a means of accumulating and analysing all the available facts about the aquaculture sector. Useful indications of the data to be collected can be found in the table of contents of existing planning documents. One example (Figure 4.5) can be found in the National Sector Reviews of African countries currently being prepared within the framework of the FAO Fisheries Department Regular Programme activities of the Inland Water Resources and Aquaculture Service; another example is the Development Plan for the State of Hawaii, prepared in 1978 (Figure 4.6).

Careful discrimination is needed in selection and rejection of data to be collected and analysed. It follows that the scope of the survey should be limited and focused on particular issues. In the following example it is assumed that the country involved is a developing one, relatively newly involved in aquaculture, and that priority is given to the formulation of a development strategy (discussed subsequently in section 4.6). It is also assumed that the survey material is partly based on field visits and detailed discussions with officials of the ministry of agriculture, the department of fisheries, the ministry of finance and planning (or their counterparts), and other governmental and private organizations involved in fisheries development, in addition to detailed study of existing censuses and studies.

Summary and conclusions

A useful introduction to the sector review is perhaps surprisingly a short statement headed 'Summary and Conclusions'. It cannot be written, of course, until the whole survey has been completed, but its preparation acts as a discipline to those preparing the survey to synthesize data relevant and directed specifically to meet objectives. It is often difficult to sort out what are important facts or factors in a report if it is written with no particular aim other than to provide information. Besides which, if the survey is to have the desired effect among legislators and senior government officials it has to attract attention. Therefore it must be written concisely in clear language.

Table of Contents from the National Sector Review, Ghana

1. General background
Physical Geography
- Relief
- Vegetation
- Geology and soils

Climatology
- Air temperature
- Water temperature
- Rainfall
- Evaporation, humidity, and wind

Hydrography and limnology
- Rivers
- Lakes and reservoirs
- Flood plains and mangroves

Land and Water Use
- Agriculture zonation
- Agricultural production
- Agricultural development potential
- Water use

Demography
Nutritional Budget
- Fish protein requirements
- Fish production requirements
- Protein cost index

Fish Marketing
- Fish prices
- Fish imports/exports

National Infrastructure
- Transport and communication
- Industrial development

Economic Status
National Food Policy and Fisheries Development Plan
- Fishery development
- Fishery training and research
- Foreign assistance for fisheries

Fisheries Development Organization

2. Fish and fishery resources
Status of the Capture Fishery
- Public sector
- Private sector

Capture Fishery Potential
- Marine capture fishery
- The freshwater capture fishery

Development Programme
- Government development policy
- Development aid

3. Aquaculture–status and development
Status of Aquaculture
- History of fish farming
- Public sector of aquaculture
- Private sector of aquaculture

Major Cultivated Species
- Indigenous fish species
- Introduced fish species

Fish Farming Zonation
Fish Culture Development
- Major constraints to development
- Extension services for fish farming
- Economic situation
- Financial support

4. Aquaculture development rationale
Site Selection
- Water resources
- Climatic conditions
- Topography and soils

Species selection
- Bioeconomical criteria
- Biotechnological criteria
- Pathological criteria
- Marketing

Process Selection
- Manpower availability
- Technological and financial assistance
- Availability of industrial inputs
- Existing infrastructure

Conclusions

5. References

6. Appendices

7. List of tables (of data)

8. List of figures (maps, charts, diagrams)

Figure 4.5 Summary of survey data for aquaculture planning (Source: Balarin, 1988).

The section, for example, may start out by stating what is the annual total and *per caput* consumption of fisheries products in the country, the proportion supplied locally, its share of the animal protein intake, and its cost in relation to other sources of animal protein. The share of aquaculture

<table>
<tr><th colspan="2">Table of Contents from the Aquaculture Plan for the State of Hawaii</th></tr>
<tr><td>Summary
1. Introduction
Opportunities for Aquaculture in Hawaii
– Constraints to development
– Aquaculture development
2. Summaries of assessments
Natural Resources
• Factors
• Maps
• Future area assessments
Technical Resources
• Effectiveness of past research
• Development of freshwater prawn farming
• Laboratory research facilities
• Outdoor research and demonstration facilities
• Extension/advisory services
• Analytical services
• Education and training
• Information transfer
• Complementary programmes
• Private research and consulting activities
Legal and Environmental Constraints
• Major legal bases
• Jurisdictional overlaps
• Common data requirements
• Importation
• Shellfish regulations
• Findings
• Conclusions
Species Selection and Assessments
• Species selection
• Individual species-group assessments (Eight reviews)
Marketing
• Seafood consumption in Hawaii
• Distribution of seafood products imported in Hawaii
• Distribution of local seafood products in Hawaii
• Market potential
• Marketing opportunities for the aquaculture industry</td><td>• Constraints to marketing cultured seafood
• Recommendations
• Markets for cultured aquatic products, 1977 & 1982
Economic Projections and Future Considerations
• Economic projections
• Future considerations
• Impact of future trends on the growth of aquaculture
3. Guidelines for development
Goal
Objectives
Policies and Recommendations
Individual objectives, rationale, policies, and actions for:
• Unified approach
• Business climate
• New economic objectives
• Excellence in R & D, education, training, and transfer
• Vocational training and formal education
• Information center
4. Implementation
Governor, Legislature, State and Counties, and University
• The Governor and legislature
• State and county agencies
• University of Hawaii
• Lead agency
5. Budget and timetable
Assumptions
Fiscal policies
Proposed budget and some considerations
6. References
7. Appendices
8. List of tables (of data)
9. List of figures (maps, charts, diagrams)</td></tr>
</table>

Figure 4.6 An aquaculture development plan (Source: Department of Planning and Economic Development, State of Hawaii, 1978).

products in total fisheries production could be included. All this highly relevant information can be given in, say, 60–75 words.

The second paragraph could explain what are the major constraint(s) to the development of fisheries. It may well be the limited potential for further expansion of capture fisheries is due to the already high level of exploitation of these fishery resources (thus pointing the way for more attention to aquaculture). A third paragraph might then give data about the current state of and potential for fish farming or aquaculture. If this kind of information is presented in a logical and lively way, the whole survey gets off to a good start.

The overall sector

First there is a review covering the general economic situation of the country and its major policies, including the broad objectives of the public investment programme. Attention is given to factors constraining economic development as these are bound to influence the growth of individual sectors. For example, the economy might well be constrained by the narrow base of export/import-substitution crops, the poor state of basic infrastructure, the weak institutional framework, shortage of foreign exchange, shortage of skilled personnel and insufficient incentives to attract or retain them, inadequate private sector involvement, and weakness of the financial and banking sector.

This could be followed by a look at the role of fisheries in the economy, covering such items as recent total catches of resources from marine and inland waters, the importance of fisheries products in the national diet, the value of fisheries landings, the number of people involved in both capture fisheries and those exploiting inland waters (mainly as a part-time activity), the importance of fisheries at the regional level, the level of fisheries imports (if any), and the extent and composition of the export trade.

This is likely to be followed by a brief analysis of the general potentials and constraints of the fisheries sector and, finally, some general comments on the role of government in the industry's development.

Fisheries background

A general review of the whole fisheries industry is likely to start with the *fisheries resources*. This will cover:

- marine resources (territorial waters and hydrology, species, biomass and potential yield of the major stocks of commercial importance, species composition of catches, fishing grounds (including maps, giving an indication of where the various types of gear are used);
- any coastal lagoon system; and
- inland resources.

This is followed by a section on current *fishing effort and catches*. This deals with the various annual landings by type of fishery, and by major species over, say, a five-year period.

A section on *infrastructure* summarizes resources such as ports and fish landing places, boatyards and maintenance yards, cold storage and ice manufacturing plants, canneries, and other processing plants.

Aquaculture background

Analysis of the aquaculture sector broadly follows the pattern used for investigating the whole fisheries background used above.

The section on *aquaculture resources* will give attention to the various aquatic species tested for aquaculture and those which have been adopted commercially. It may be, for example, that culture of shellfish and crustaceans is not practised, although traditional culture of freshwater clams may once have been an important industry. Culture of other species, including mangrove oysters, clams, freshwater shrimps, and blue crabs, etc., may have been contemplated but not yet attempted.

The status of aquaculture will describe current farming activity, production levels and efficiencies in culture, brackishwater aquaculture, cage culture, culture-based fisheries, other forms of traditional farming, the producers (full and seasonal part-time workers), an economic evaluation which briefly examines trends in capital investment, operating costs, and income on representative farms and their internal rates of return (as a useful guide to indicating farmers' ability to generate a sufficient cash flow to obtain credit for expansion, and to make loan repayments within a specified period), hatcheries and seed production units, and finally the availability of essential equipment.

Marketing, supply, demand, and prices

Artisanal processing and trade is a section which provides a description of how fisheries products are handled, processed, and sold. With regard to farming, for example, the cultured products may usually be sold by farmers direct to the consumer after harvesting (describing relevant methods) to serve traders whenever they come to buy. Usually it is found that the traditional system of product handling, processing, and marketing is deeply rooted in the country's social structure and works perfectly well (and if not, then its faults should be identified). The system is likely to be very flexible and capable of absorbing considerable quantities of fisheries products.

A section on *supply, demand, and prices* relates fisheries catches to demand, giving attention not only to nutritional requirements but to economic demand, determined by the purchasing power of the consumers, and other economic factors such as production and distribution costs. It may be found, for example, that there is not sufficient purchasing power to satisfy nutritional requirements at the current level of prices even if more products were available. In fact, with fisheries products distributed by a relatively large number of small traders/producers, and prices at their optimum (competitive) level, it is likely that prices may have to fall sharply if a larger volume of any product is to be bought by the consumer (with, for example, a price elasticity of demand close to (say) what may have been identified in the section, 'Summary and conclusions'. The export/import balance has also to be examined carefully.

Institutional aspects

This section covers a wide variety of activities which influence the general efficiency and effectiveness of the fisheries sector. Even though the interest is mainly in aquaculture, it is common to look at all aspects of the fishery sector first, and then to examine the special conditions in aquaculture. The marine and inland fisheries use many common government facilities and services.

The *fisheries administration* is required to outline briefly its functions, system of management, and staffing. Typically it will have several branches, each headed by a deputy director covering:

- operations and services,
- marine fisheries,
- inland fisheries,
- research,
- industry and utilization (such as surveillance and monitoring fisheries stocks, collection and analysis of fishery statistics, and development of appropriate fishing gear).

The survey is likely to attempt some evaluation of the performance of this administration and the position of aquaculture within it. Current aquaculture legislation, particularly in relation to land and water rights, should be reviewed.

Producers' organizations will provide details about the main objectives, membership, operations, and methods of management.

Those agencies involved in *research, extension, and training* will account for all programmes underway in these three important areas, their cost effectiveness, relevance to the industry's problems, as well as the respective organizations involved.

Similarly those responsible for *other matters*, such as the agricultural development banks, will provide a review of credit facilities available to aquaculture farmers, and their recent experience of debt repayment.

Current fishery and aquaculture policy and programmes

Fishery policy looks first at policy objectives even if there is no written document setting them out. For example, these may be (in general) to maintain the catches of the:

- artisanal fishery, as a low cost source of animal protein and a source of income and employment for the artisanal fishing community,

- industrial and semi-industrial fisheries, as a source of food, income and foreign exchange, and
- inland fisheries, as a source of animal protein and employment in the hinterland.

Specific objectives are likely to include improving the government's management capability for all aspects of the fisheries, especially by the implementation of an effective monitoring, control, and surveillance system, as well as the capability to provide more effective extension services for aquaculture, and improved credit availability.

The survey team should not only describe all these activities but also make elementary appraisals of their effective implementation, and to examine also legislation and regulations which have direct relationship to aquaculture, and to both marine and inland fisheries.

Current aquaculture policies, assuming that the government has already included targets for the development of aquaculture in its national agricultural development plan, can be compared with actual performance. Quite probably some targets proved to be rather optimistic, mainly due to lack of adequate funding, but some kind of appraisal is useful to the 'diagnostic survey',

Finally, the *extent of funds received from external assistance*, as well as those committed for the immediate future, should be accounted for – probably to support activities in production, preservation, post-harvest losses, and training in aquaculture.

Analysis of the development constraints

Once the first stocktaking part of the survey has been completed, the way is open to make the diagnosis – to decide what is holding up the rate of development in the aquaculture sector. If the assessment is reasonably good, the diagnosis may not indicate any drastic treatment is needed. However, it may be otherwise. Each of the main factors influencing the performance of the industry needs to be examined in turn. In both aquaculture and inland fisheries, for example, there will be an examination of available resources and productivity (partly relating to the levels of exploitation), of the economic performance of the various systems and practices, such as intensive or semi-intensive systems, pond farming, cage farming, hatchery production, etc. Sector infrastructures will be assessed as well as distribution. Private sector activities will come up for review, including competitiveness, and levels of efficiency.

There will undoubtedly be institutional problems as well; for example, possibly neither the existing nor the proposed revised law(s) offers the necessary legal framework for certain types of farming, or contains regulations (regarding cage size, or water restrictions,) which are not sufficiently enforced. There may be inadequate coordination between the various government agencies concerned with aquaculture and fisheries, especially in the areas of external assistance and research, while the dialogue between the government and representatives of the industry may be far from constructive.

As to aquaculture, the survey may have identified some difficulties relating to different practices of culture. In the development of extensive culture-based fisheries, for example, the main factors limiting its success may include: inadequate fry production and distribution capability; poor knowledge of locations and communities most likely and able to produce benefits; appropriate harvesting methods probably not adequately known or organized amongst target communities; no technical knowledge of culture-based technology in terms of nursery, stocking, harvesting, and engineering requirements. Where aquaculture is practised on irrigation schemes its limited success may be caused by: supplies of raw materials being too expensive or inadequate; inappropriate technology; cultural attitudes in poor rural areas more aligned to crops than to livestock; extension programmes have probably been mounted on too small a scale, etc. This kind of analysis is invaluable when it comes to developing a strategy which sets out goals and priorities.

A typical strategic approach might have the following five elements:

- to ensure that an adequate basis of technical knowledge is available both at senior advisory level and at regional technical assistant level and that there is adequate manpower located in target development areas;
- that there are properly supervised credit schemes available with appropriate loan terms;
- that adequate facilities exist for the production and distribution of seedstock;
- that adequate information exists on distribution and status of surface and ground waters, in order to plan and evaluate aquatic farming and enhancement activities. In particular, there may be a need for up-to-date information on the numbers, distribution, and relative permanency of small temporary dams and fish plains in the specified regions;
- more effective efforts may need to be used to develop improved technologies in polyculture, harvesting frequency, feed mixes, seed production, etc. In particular, research and development programmes at specific institutes and universities may need to be coordinated toward common practical goals, with needs clearly stated at producer level.

This whole exercise provides the basis for the action (and investment) programme to be recommended for the aquaculture sector.

4.6 Formulation of development strategies

The definition of a development strategy

A development strategy provides a frame of reference to facilitate identification of relevant policy alternatives, and the choice of those likely to be most effective. By laying out the broad approach or general direction to be taken to achieve a specific objective or target, a strategy allows planners to choose from a variety of potentially usable policies, projects, programmes, and other measures, those which are best suited to that approach, and to integrate them with each other.

As the link between an objective and the policies required to achieve it, a strategy is a convenient means for testing whether all the policies and programmes needed to realize an objective, are being or can be adopted.

A strategy is also a means of selecting elements or factors on which to concentrate in developing the sector. These elements involve choices among such things as the different production technologies, spatial arrangements, time horizons, production mixes, and human groupings, etc., as well as the emphasis and sequence to be given to each in achieving a given objective or target.

A development strategy may thus be seen as a framework for a consistent set of policies, measures, and activities, within defined constraints, which together constitute an organized attempt toward fulfilling a particular objective or target. However, the adoption of a strategy to achieve a given objective or target does not imply that everything outside the strategy will be ignored – only that the strategy will be emphasized; for example, adoption of a strategy to improve self-sufficiency in fish production would not mean that export expansion would get little or no attention, but that self-sufficiency would get greater attention.

Development strategies in theory and practice

The theory of strategies has been described. In practice, however, many policies exist (in national as well as in aquaculture planning) without any reference to a clearly defined development strategy. The absence of a policy framework is a conspicuous feature of several countries with 'planned' economies, and in spite of the fact that they are well endowed with planners. Many poor countries, in practice, pay little attention to the development strategies required to achieve their objectives, and several have policies headed in different, and often opposite, directions.

A Thematic Evaluation of Aquaculture, carried out by UNDP, FAO, and the government of Norway (FAO, 1987) studied 53 international assistance projects, several of which were directed towards planning development of the aquaculture sector. The study found that the contents of the plans, in most instances, produced through technical assistance were mostly generalizations regarding goals, and without any realistic regard to the efforts required for the tasks, or the low levels of available financial, human, and technical resources. Furthermore, it found it was unusual for governments to have an appropriate strategy for aquaculture development, and even when a strategy existed planning was still carried out without knowledge of the skilled manpower and supporting resources which the government was willing to invest. Production targets were often set with little indication of how the necessary inputs and infrastructure were to be provided. In general, there was a preoccupation among 'planners' with what was technically possible rather than with what was economically possible and socially acceptable. These weaknesses were made worse by the common tendency not to modify the plan (or strategy) in the light of changes in the availability of resources.

Planning for aquaculture development has been weak

The evaluation of 53 international assistance projects concluded that:

> 'Only 13 projects produced outputs related to planning. This is less than the number originally meant to include planning at the national level. Although the outputs produced are in the form of planning documents, close examination shows that almost all of them are a mixture of plans and policy statements. There is only incomplete evidence of the extent to which these documents have been useful to governments. In many instances, the contents of plans produced through technical assistance were weak. The general standards of documented aquaculture development plans are poor and are often not rationalized in terms of government's aquaculture policy objectives. While production targets are set (at times, for periods even longer than the conventional five-year development plan), the corresponding indications of how the necessary inputs and infrastructure are to be provided are lacking.'

Source: FAO, 1987.

The problem with many aquaculture development plans in particular is that they fail to distinguish between objectives, strategies, and policies. The distinctions have already been described (see section 4.3). Consequently, when an objective is taken to be the same as a strategy, the need for a unifying framework for the policies is likely to be overlooked; and when a policy is taken as a strategy, it is likely to be too narrowly defined to provide that unifying system.

A well defined strategy allows resources to be combined in specific ways to achieve an objective or target. Clear definition is important because different strategies use resources in different ways; for example, one strategy may rely on a discrete and significant technological advance (such as curing a disease), while another may require gradual improvements with traditional inputs.

Types of strategies

Every country selects its own strategy for developing an aquaculture industry according to its own individual circumstances. Therefore, it is not possible to provide an overall prescription or model. Primarily, attention is usually directed to the 'supply side', but equally important are policies on the 'demand side' to ensure that the population, particularly the poor and malnourished, have secure access to food at prices they can afford.

A government can play a major role in removing food insecurity from such people (especially women, children, and those living in backward rural areas). In these circumstances, the strategy adopted might be based largely on extensive and subsistence aquaculture, characterized by its compatibility with other key rural activities, production of small quantities by large numbers of small-scale farmers, and little change in existing distribution of land and water rights (see Chapter 1, section 1.2). Alternatively, a strategy could favour semi-intensive aquaculture by a few entrepreneurial farmers with larger facilities, as they may have a better chance of overcoming constraints and are often willing to try new technologies, but who also still need assistance to assure them of continued and exclusive access to culture sites and reliable information on markets.

The contrast between these two strategies largely concerns the 'concentration versus the dispersal' of resources. If the objective is greater equality in the distribution of rural incomes then the first strategy outlined is more acceptable; if it is a rapid increase in output, without regard to income distribution or employment creation, then the second is preferable.

Because of the possible high social costs of a strategy which concentrates resources, preference is frequently given to a strategy of 'gradualism'. Experience suggests that the gradualist approach is the most stable and reliable way to develop aquaculture in poor countries, although it may not be the quickest. The less urgent the need for short-term increases in output, the greater the benefits of a strategy which combines limited amounts of capital with increased use of surplus labour to improve production methods gradually on as many farms as possible. The gradualist approach has shown that not only can it increase output but it also provides more employment and a more equitable distribution of income than one based on concentration of resources.

The choice of the right strategy often revolves around the degree of production intensity, as this greatly influences the utilization of available resources. The three principal aquaculture production systems, namely extensive, semi-intensive, and intensive, and their differences, have all been described in detail earlier (see section 1.4). Similarly, the availability of culture species and variations in the culture practices based on these three systems have been described, together with the potentials for enhancement (see section 1.3), and these all may weight any choice.

Some strategies require a longer time to carry out than others. It is therefore desirable to determine how long a given strategy will take to produce results. If quick results are wanted, then a strategy which relies on major changes in technology, administration, or organization should be avoided. Long lead-times are typical in aquaculture, so that careful evaluation strategies by their short- or long-term effects. If both types are attempted at the same time they usually conflict with each other. Other strategies may favour producing species for import-substitution, the expansion of traditional exports, or to produce new exports (such as marine shrimp).

Aquaculture may also take part in a strategy of agricultural diversification. This uses the principle of 'opportunity cost'. For example, marginal land and water resources, which previously had been neglected but which, with relatively little investment, can be put into aquaculture. This expansion in agricultural output can be relatively cheap, as output of agriculture (in the narrow definition) is not likely to be lowered. On the contrary, it is more likely to increase, as higher levels of income from diversification provide the basis for more investment and higher levels of everyday spending.

Who selects strategies

It is clear that the process of selecting a strategy, like that of identifying development objectives, involves decisions which will benefit one group of producers and penalize another. A strategy may be 'neutral' in theory, with regard to scale of operation or location, in the sense that any producer anywhere can apply the package of inputs required to obtain profitable returns. But in practice neutrality breaks down quickly. Larger (and wealthier) producers, with better access to credit, inputs, and extension advice than small (and poorer) producers, benefit more from most strategies.

Selecting a strategy is more than a simple technical issue. Although a strategy has technical content and implications which planners and other technicians are qualified to deal with, the political and social aspects of strategy are even more important. Consequently, it is the political leadership, assisted by planners and other technicians, which has the ultimate obligation and responsibility for selecting the strategy to be followed.

Constraints on the choice of a strategy

The need for systematic analysis. Because ideals of social equity and economic efficiency often conflict, a strategy for the aquaculture sector must aim at an acceptable compromise between conflicting social and economic aims, and obtain efficient use of the scarcest resources. This provides the basis for systematic policy analysis of, for example, the conflict for primary resources needed by aquaculture, agriculture (both crop and livestock husbandry), light and heavy industry, tourism, and urban development (see Chapter 2). Setting priorities between these competing sectors is the crux of the national planning problem.

There is also the need to examine the interface between aquaculture and traditional capture fisheries. This has some conflicting and some complementary relationships, as well as positive and negative effects of aquaculture on the natural environment. Interaction between a proposed aquaculture project and the natural environment can be sufficiently complex or pervasive to require detailed analysis. Fortunately this is now much better understood, but skilled technical experts, working in teams, are required to understand the environmental system, its physical and energy flows, and its modification by human intervention. They must appraise the potential weaknesses and susceptibility to degradation of the ecosystem by the intervention of aquaculture (or other human activities). Their analysis should identify the renewable and non-renewable natural resources which will be used by the project, and determine the ecological consequences. It should also identify the most sensitive points in the ecosystem which the project could influence adversely, appraise pollution risks, and generally assess the extent to which aquatic farming and production/processing systems will be ecologically sustainable. Parts of such a project may be specifically designed to benefit the environmental issues, such as water quality schemes which ameliorate already polluted waters. Proper analysis requires the collection and consideration of as much information as possible about the costs, benefits, advantages, and disadvantages of each alternative, as well as a determination of the relative abundance and scarcity of major resources.

While benefits and advantages of a possible strategy are generally easy to assess, its full costs and disadvantages are frequently less apparent. One reason is that the nature of constraints for a given strategy may vary greatly from one country, or one region, to another. A second reason is that it is necessary to adjust a strategy to the special situation in a country before all costs and disadvantages can be determined. Planners frequently fail to recognize that substantial differences in constraints, from one country to another, make it essential to identify them in specific environments, rather than to look for a strategy suitable in a wide range of conditions.

Resource constraints. Recognition of resource constraints influencing the choice of an aquacultural strategy is frequently the difference between success and failure in achieving targets and objectives. One of these is imposed by the natural conditions in a country. Soils, climate, endemic disease, and scarcity or abundance of water, are examples of natural conditions which set limits on aquacultural strategies. Planners in India, for example, must take cognizance of monsoons, just as those in Israel need to allocate scarce water to high value fish production. However, resource constraints are not always permanent, and things may change through new technologies. Therefore planners must look ahead and try to estimate which resources are likely to get scarcer, and which more plentiful.

A second constraint is a country's stage of aquaculture development. Clearly, in the early stage a country should not select a strategy which, like mechanization, requires skills, foreign exchange, and other resources which it may lack. A well-advanced country should not adopt a strategy which emphasizes labour-intensive techniques. Any strategy must be appropriate to the actual situation in a country.

A country's social, political, and economic institutions and beliefs are, collectively, a third determinant of strategies. At the international level they can embody attitudes toward the role of foreign trade, aid and private investment, the importance of inter-regional cooperation, and the desirability of attaining economic independence. At the national level they can reflect public attitude toward such things as the roles of the private and public sectors, regional development, family planning, unemployment (and therefore, among other things, the choice of technology for public investment projects), the pattern of income distribution, levels of consumption, retention of traditional ways and forms, and institutional reform.

Some constraints are directly self-imposed, such as a government's attitude toward the private sector, or its price policy on foodstuffs for urban workers, which may actually discourage production. Others are more indirect; for example, foreign exchange constraints on the choice of a strategy are usually independent of a country's social, political, and economic attitudes. However, if a country avoids certain kinds of foreign aid as a matter of principle, the foreign exchange constraint is at least partly self-imposed.

Planners may or may not have the power to change a government's self-imposed constraints. For planning purposes they should determine the extent to which they must accept, as given, existing attitudes and institutional constraints on the choice of strategies, and the extent to which the constraints are subject to revision. Only a good understanding of the thinking of the political leadership can make it possible for planners to help select an appropriate strategy.

In many countries the state of public administration is a serious constraint to aquacultural development. Failure to recognize this has led to the adoption of strategies which make demands on public administration which are heavier than it can support. A government administration is often called upon to carry out more projects, or to intervene more frequently, than it can manage effectively.

In choosing a strategy it is important to distinguish between constraints at the national, sector, and enterprise levels. Lack of investment resources may be a constraint at the level of the aquacultural sector but not at the national level, where a government has decided to concentrate all available resources on industrial development, rather than agriculture (including aquaculture). Similarly, just as constraints may differ between the national and sectoral levels, they may differ between those at either of these levels and with the enterprise level. For example, availability of fertilizer or feed may not appear a constraint at the sector level, but unreliable deliveries at the enterprise level may be a serious problem; and while skilled manpower may be a limiting factor at the national level, farmers may have all the knowledge required to produce more if only deliveries of seed and feed could be improved.

Because constraints faced by producers actually retard development at any given time, assumptions must fit the difficulties which they face if effective strategies are to be selected. Unfortunately, these answers are often based on the wrong questions, as the assumptions were applied to constraints at national or sectoral levels, instead of those at the local level. Therefore, a strategy must be designed to ensure that producers see the actions to be taken as contributions to solving their problems and achievement of their objectives.

Sometimes a single strategy will not be enough. This is true, for example, where improvement of both inland fishing and aquatic farming is attempted. As the problems and constraints of inland fisheries differ from those of traditional aquaculture, two strategies, or two distinct components within an overall strategy, have to be identified. One will deal with enhancement of commercial inland fisheries, the other with aquaculture.

In some countries a single strategy may not be possible for reasons of the diversity of ecological, climatic, or economic regions. In Brazil, for example, great regional diversity, and a tradition of regional autonomy in matters of development, make the formulation and execution of unified aquaculture strategy impracticable. Turkey has three distinctly different regions – the highland plateau and two coastal zones – each of which requires a development strategy different from the other. Most countries in fact have the division of inland freshwater regions and coastal brackishwater or marine regions, which may be complicated further by large lagoons or mangrove swamps which require their own particular attention.

Frequently differences in development strategies arise from economic differences between regions of a country. Thus economically backward regions require development strategies of their own and differ from those of other parts of these countries. The same may be true for a region which is far more advanced than others.

The best strategy is one which is least bound by constraints. Thus, in a country with a foreign exchange constraint, a strategy which makes maximum use of domestic resources (like labour) is better than one which requires costly imports (like mechanization). Implicit in strategy selection is the requirement that the greatest use be made of a country's abundant resources.

While constraints impose limitations on the choice of strategy, they can be dealt with either by sidestepping them or devising a strategy with elements which fit the constraints.

Implications of a strategy

Before selecting a strategy it is essential to explore all its possible repercussions. Sometimes it may be easy to see that a certain strategy will not achieve its objective. For example, a strategy which emphasizes sophisticated and capital-intensive production techniques is unlikely to raise the incomes of poor fisherfolk. On the other hand, even if it is clear that a strategy will succeed there may be side-effects which need to be considered. The process by which a strategy is selected from among available alternatives should, therefore, include careful consideration of all possible negative effects. Some negative effects might have been foreseen and provided for with contingency measures if a more systematic analysis of the strategy had been made prior to its adoption.

Finally, it is vital that those closely affected by a strategy are kept fully informed about its nature and implications. Too often the one certain quality of a strategy is its lack of clarity. Even if this is not a problem, a strategy may not encourage much positive response and support because of failure to keep producers and local officials fully informed.

Some country strategies

While developing countries have little difficulty in identifying objectives for their aquaculture development plans, few have reached the point where the strategies for attaining these objectives are consistent and defensible. For example, in sector plans of many countries of eastern Asia, it is clear that the emphasis is shifting from traditional marine capture fisheries to inland fisheries and aquaculture. This is a result of overfishing and diminishing resources in the major fishing grounds, and the high cost of fuel.

In Japan, a country with a long tradition of fishing, a new strategy has been carefully defined to meet the high national demand. Increased fish production now involves the enhancement of diminishing stocks of commercially valuable species by restocking large quantities of seed in open waters through artificial propagation, intermediary nursing, expansion of natural nursery grounds by seaweed plantation and the construction of artificial fish reefs. This strategy for increasing 'culture-based fisheries' relies on aquaculture methods and techniques for the production of seed for stocking coastal and marine waters, and their harvest by coastal fishermen.

The government of Vietnam is giving importance to fish and shrimp farming as an integral part of agriculture. It is particularly interested in the introduction of modern technology to its traditional fish culture.

China has followed a strategy since about 1979 to bring about the rational utilization and management of water and land resources, and particularly its coastal waters. It has focused on aquaculture development. The government of China regards the importation of techniques, engineering expertise, and equipment for the exploitation and utilization of its fishery resources, as major components of its modernization plan. Various policies have been formulated redirecting emphasis toward the development of marine and freshwater aquaculture, providing considerable financial, material, and technical support to ensure attainment of production goals. This whole programme has been so well formulated and implemented that output in 1991 was double the level of 1980 (4.5 million t) and well ahead of target. China's projected target for the year 2000 is 18 million t.

Finally, it is useful and interesting to look at a proposed strategy for Canadian aquaculture on the threshold of commercial development. Many of its proposals, while directed to a comparatively wealthy country, have some relevance to poorer countries. The strategy was prepared by the Science Council of Canada in 1985.

National strategies should be cogently stated and well-publicized

The proposed strategy for Canada states:
'It is recommended that federal and provincial governments should:

- recognize commercial aquaculture as a high-priority opportunity for development;
- establish lead agencies to coordinate and encourage development of this emerging science-based industry;
- direct these agencies to coordinate funding from all sources and, with the advice of Regional Aquaculture Coordinating Committees, channel funds into development of the species identified as having the greatest commercial potential in each region;
- cooperate closely with the industry to support the creation of private-sector hatcheries, sufficient to supply the needs of a rapidly growing industry with superior strains of seed stock;
- extend assistance to the industry for an initial five-year period to assist creation of effective producer organizations to provide a forum of continuing consultations with government, to cooperate and share knowledge within the industry, and to enable producers to develop self-discipline concerning quality assurance standards and market promotion;
- work with the private sector and with representatives of native groups to prepare enabling legislation to establish a legal framework for the conduct of commercial aquaculture, to clarify or create the necessary rights, supply financial incentives and regulate the industry as appropriate;
- support development of commercial aquaculture with retraining programmes, local employment programmes and community development projects;
- cooperate to develop centres of excellence to serve the needs of the aquaculture industry;

- clarify departmental mandates regarding aquaculture and evaluate programmes with a view to enhancing support for aquaculture; in particular, Investment Canada should use its new proactive mandate to seek out sources of domestic and foreign capital with a view to facilitating investments that will stimulate the development of Canadian aquaculture capabilities; and External Affairs, through its science counselor network, should assist entrepreneurs to identify opportunities for international technical exchange agreements;
- develop policies integrating the needs of fish population enhancement with the needs of a commercial aquaculture industry;
- establish appropriate services for the industry, particularly veterinary and pathology services, in consultation with the industry.

The industry should:

- create effective producer organizations to provide a forum for continuing consultations with government, to cooperate and share knowledge within the industry, and enable producers to develop self-discipline concerning quality assurance standards and market promotion;
- seize opportunities for technology transfer and develop new markets for species under commercial production in Canada;
- create appropriate conditions to attract investment capital by:
  - participating in technical exchange agreements and adapting foreign exchange and domestic best-practice technology;
  - examining ways of reducing risk to meet standards set for aquaculture insurance;
  - raising the quality of their business planning;
- develop commercial applications for Canadian raw fish wastes and promote scientific research, particularly in fields such as fish nutrition and feed technology.'

Source: Science Council of Canada, 1985.

The key role of strategy in planning for aquaculture development is clearly brought out in India's Seventh Five-year Plan (1985–90). After reviewing development in the Sixth Plan period, the text moves straight to 'Approach Strategy and Programmes', with reasoned discussion justifying the choice of measures to be adopted. They are based on a strategy of intensive fish farming in tanks and ponds through the Fish Farmers' Development Agencies, and the introduction of prawn farming in brackishwater areas by establishing 'Area Development' prawn-farming estates. The Central Institute of Coastal Engineers for Fisheries will intensify its efforts to accelerate the brackishwater survey and the design of fish farms. These culturable resources will be the only source of additional inland-fish production during the Seventh Plan period. This approach will also create a 'cadre' of trained fishfarmers to undertake scientific fish farming. The next section of the plan deals with targets for fish production and the details of selected development programmes (including some administrative changes). Every attempt will be made, for example, by all states to follow West Bengal's achievement of becoming self-sufficient in fish-seed production. Without its clearly stated strategy, this plan would lack the framework against which its policies could be judged.

PART II
AQUACULTURE SECTOR MANAGEMENT

Chapter 5
Organization and Management of the Sector

Synopsis

The chapter notes the wide diversity of development of individual countries, and the inputs and support needs for successful development of the sector. The different responsibilities of the public sector are first discussed in terms of identifying objectives, establishing policies and selecting policy instruments, and other possible activities. This is followed by the position of the aquaculture administration and other institutions, such as parastatals, in the government. The last section examines some of the practical issues which arise in establishing or modifying an organizational and management structure for the sector, and the tools available to the government for management. In all these sections, examples of various government infrastructures and parastatals are given, with comments to point out advantages and disadvantages.

5.1 The role of the public sector

5.1.1 Background

The practices of aquaculture in many countries can be traced back through the centuries, but the present concepts of organization and management only originated within the last hundred years. Even then, this comparatively short period does not provide a precise perspective of the industry. Early steps were tentative and exploratory, and mainly concerned with formalizing and, where necessary, regulating traditional practices, usually in relation to legal rights.

Modern aquaculture systems and practices have burgeoned in the last 25 years, and development has acquired new scientific and particularly economic dimensions (see Figure 5.1). These have all but revolutionized traditional concepts, and expanded the sector's potential out of all previous expectations. Naturally this technological progress has been much more rapid in certain countries and regions than in others, and for very different non-technical reasons. This has resulted in a variety of organizational

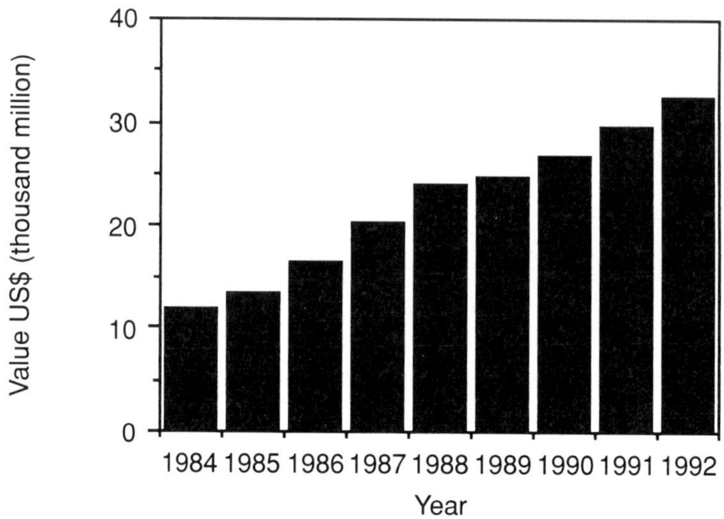

Figure 5.1 Global value of aquaculture production, 1984–1992 (In US$ thousand million) (Source: FAO, 1994).

structures in control, with a range of different policies and policy instruments to manage their respective sectors. In the USA, for example, organization in aquatic farming has been built around its structure for agriculture. With the heavy investment in research by the private sector, the government added its financial support through a new version of the old land-grant system. This was responsible for the successful development of American agriculture in the first half of the century, linking selected universities for research support with a network of agricultural extension agents throughout the country to participating producers backed by incentives. On the other hand, enhancement of the indigenous migratory and sports fisheries through aquaculture continued to be organized entirely around the public sector, primarily as part of federal mitigation projects for changes caused by large water and forestry exploitation.

In Scandinavia, Norway has built its sector around the organization of its vital fisheries industry. The government supported the new industry through research and marketing primarily to compensate for a vanishing fishing industry. Like Sweden and Finland, Norway knew it had to sustain the integrity and economic base of remote coastal fishing communities. In Europe, England and Scotland made aquaculture one of several priorities in new parastatal organizations for fisheries development in general and the economic development of the highlands and islands, respectively. France, on the other hand, simply made it an extension of a century-old tradition of coastal farming and fishing through cooperatives and fishermen's associations, and backed them with a strong emphasis of organization and support in research by the government.

All this was in sharp contrast to the tropical and sub-tropical countries of East and Southeast Asia, where there was a long history of aquaculture without much organization. Therefore the emphasis continued to be '*laissez-faire*'. The general entrepreneurship of the Asians, particularly the more advanced nations such as Japan and Taiwan, made them quick to apply modern scientific techniques to take advantage of the increasing wealth and expanding market opportunities in the region, which stimulated rapid growth. Their enterprise was soon followed by the private sector in Indonesia, Malaysia, the Philippines, and Thailand. Consequently, the expansion of modern aquaculture in most of Asia was so fast that it outstripped the supporting infrastructure. This led to uncontrolled development, with all the attendant problems of indiscriminate land-use development, political confrontations over water and land rights, and public health problems in respect of both products and pollution side-effects.

Where there was no strong interest in fish at all, the situation was very different. In countries of South and Central America, where there was little dependency on inland fisheries and no tradition of aquaculture, organization was almost a 'catch-up' reaction to the private sector introducing and applying modern culture techniques for farming species taken in large numbers from the wild. In Africa, where inland fisheries and hunting contributed to the basic animal protein in the diet of the people, organization was built into rural development programmes with subsistence farming as the principal method of production usually under the guidance of the ex-colonial fisheries department.

Nonetheless, in spite of the rapid world-wide economic development of aquaculture in recent years, its relative importance within the overall

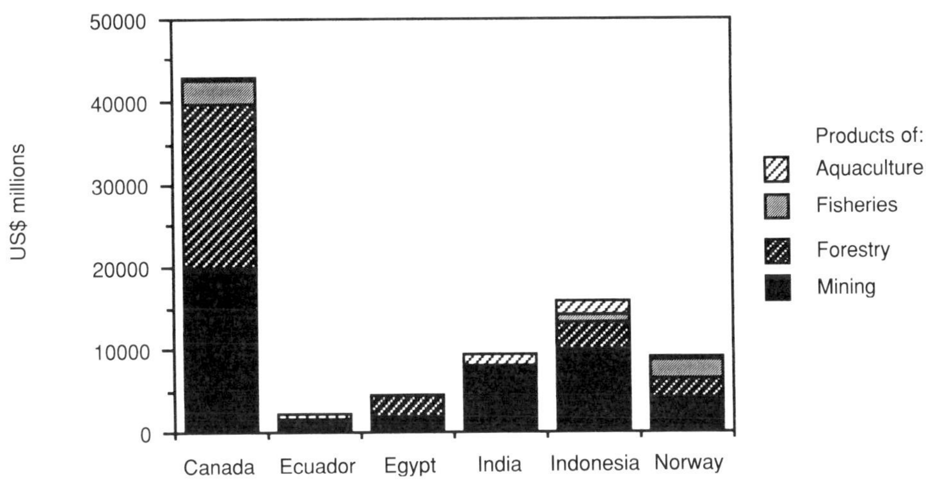

Figure 5.2 Some national values of aquaculture production compared with trade (exports and imports) of other sectors: 1989–90 data (values in US$ millions).

national socio-economic pattern has a long way to go before it has any effective political leverage in the competition for increased support services to accelerate development further (see Figure 5.2).

One reason for the relative weakness and unimportance of aquaculture in most national economies has been the tight government control, frequently constrained by historical precedents. Although the situation in Hungary is in some ways unique, as the entire fisheries production is freshwater and the country had until recently a centrally planned economy, it typifies the constraints faced by many producers, in addition to added financial burdens from taxes levied on sales of production and seed, as well as on purchase of equipment, oil and lubricants, and other farm inputs. By comparison, in most countries of West Asia, aquaculture is encouraged and licences are not required for utilizing private property, and permits and lease rights are only required for use of government-owned water bodies and adjacent lands. In addition, subsidies in the form of cash and kind, ranging from 10–50% of project cost, are available to fish farmers in some countries, such as Bangladesh, India, Nepal and Thailand.

Historical precedents have constrained aquaculture growth in many countries

Traditional fishing in land-locked Hungary lacked any legal protection for fishermen *vis-à-vis* the claims of aristocratic landowners. Water laws passed in 1888 provided certain safeguards, but proved inadequate in that the rich could outbid traditional fishermen for licences or permits to fish more productive areas. This led to misuse and over-exploitation of such waters, further legislation, and finally state intervention when leasehold tenancies were replaced by cooperatives, and in certain instances by state farms with specific research and development objectives. The present health of the sector in Hungary is reflected by the fact that consumption of fish has quadrupled since 1950, compared with only the doubling of meat consumption. Even so, fish consumption only accounts for 1% of total food expenditure.

5.1.2 Responsibilities of the public sector

Identifying objectives for the sector

The diversity of their backgrounds, and the equally wide variation in their stages of relative development, are reflected in the present government organization in each country, and the variety of long and short-term objectives appearing in national development plans and government policy statements.

Fortunately, emerging sectors have been able to find several places where they can fulfil one or more of the many economic and social objectives identified by their respective governments, and therefore garner support.

National objectives for aquaculture are often associated with economic and social objectives

Typical long-term objectives are:

- national food security,
- management of both renewable and non-renewable resources,
- protecting the aquatic and its interdependent environments,
- replacing capture fisheries with aquaculture products.

Short-term objectives, more appropriately, include:

- increasing national food supplies,
- increasing foreign exchange earnings,
- creating employment, particularly in rural areas where other opportunities are limited,
- maximizing farmers' incomes.

Establishing policies for the sector, and selecting appropriate policy instruments

The impact of the large selection of objectives in itself influences the policies and selection of policy instruments for managing the sector, as each calls for its own specific inputs. Some of these, in turn, signify the relative level (both high and low) of political importance given to the agriculture and fisheries sectors generally. This variation is due to the 'philosophy' of each individual government. Philosophies range from the centrally-planned economic approach, characteristic of the old Eastern bloc countries of Europe, to the free-market economies of Western Europe, USA, and many Asian countries; but they also range from non-intervention and self-discipline in the industry, characteristic of Japan and some Asian countries, to an over-cumbersome framework of bureaucratic regulation and direction, as inflicted by the European Union (EU) on its member States, and the federal and state authorities in the USA.

In general, public administrations and agencies have tried seriously to fulfil their organization and management responsibilities by providing their emerging aquaculture industries with policies and supporting policy instruments to achieve specific objectives and create an environment favourable to growth.

Public administrations have responsibility to provide the sector with a regulatory framework and supporting services

Priority areas of responsibility include:

(1) *Legislative and regulatory responsibilities*, such as:
- formulation of national development policies,
- legislative protection of production,
- public health controls over aquaculture products and by-products (including pollution controls where applicable),
- control over the introduction of non-indigenous species, and of disease in cultured species,
- conservation of the environment and its coherent use,

(2) *Fiscal responsibilities*, such as:
- ensuring adequate credit and other financial services,

(3) *Support services*, such as:
- scientific and technological research.

Arising from the primary responsibilities, governments have been called upon to provide ancillary support in almost every aspect of sector operations. These are summarized in Figure 5.3. Whether governments have undertaken these responsibilities directly, or simply ensured that they were available and/or undertaken, has been a matter of choice. In the Philippines, for example, extension for the new marine shrimp industry has largely been provided by private sector consultants, rather than the government's large extension service which has supported the traditional fish farming industry for three decades.

Delegation of responsibilities

Many national or central governments have found it more effective and efficient to delegate part or all of their responsibilities to provincial, state, or even more local authorities. They have, however, invariably retained certain overriding powers in case they were needed in exceptional circumstances, or emergencies.

Few countries accord any degree of political importance to the fisheries sector, and therefore the aquaculture sub-sector even less. Consequently, there has been considerable benefit from the concept of delegating aquaculture responsibilities to provincial and local levels, as the closer the authority has been to the actual operational activities the better has been the understanding and empathy for the needs of those involved. Furthermore, the authorities have then been more readily accountable.

In many countries, most of the major public sector responsibilities in respect of aquaculture were already in place within the national infrastruc-

An illustrative breakdown against which may be listed those agencies (government and non-government) responsible for, or providing services in, the underneath areas, at different administrative levels (national, regional, provincial, district, local).

Aquaculture Development	*Farm Inputs*
population matters migration employment issues rural works programmes non-rural farm occupations rural industries socio-economic surveys statistics planning policy, programme formulation regional development manpower planning and strategy	feed supply fertilizer supply pesticides herbicides algicides seed production and distribution farm machinery and equipment broodstock resources credit farm labour water supply
Institutions, Organizations	*Farm Outputs*
rural people's organizations farmers' organization cooperative societies rural women and youth organizations aquaculture research and experimentation in: – livestock – farm management – farming systems – economics – mechanization – labor intensity, etc. aquaculture education and training in: – aquaculture – animal husbandry – veterinary science/pathology – genetics – engineering – social sciences and economics – related fields (agriculture), etc. extension or technical services to: – producers – hatchery operators – livestock handlers – farm owners, etc.	product control marketing of: – food products – non-food products – export products processing packaging storage facilities *Regulatory* land laws water laws land reform water rights conservation and eco-management water/flood control drainage and irrigation disease control livestock health and movement control food quality control aquaculture statistics and returns fiscal control

Figure 5.3 Areas of responsibility in the aquaculture sector (Source: Nash, 1992 (adapted from Rowat, 1989)).

ture. These included, for example, the regulatory controls over the production and marketing of comparable food products. Consequently, all that has been required are some minor amendments to regulations to ensure that aquaculture products were covered, retraining of the staff, and some increased funding.

Many countries have delegated responsibilities for servicing the sector to different levels

In *Pakistan*, fisheries development is decentralized and considered to be a provincial responsibility, within the guidelines laid down in the National Fisheries Policy. The Pakistan Agricultural Research Council, Federal Water and Power Development Authority, National Institute of Oceanography, Pakistan Council of Scientific and Industrial Research, Zoological Survey Department, and various universities also contribute to fisheries development.

In *India* fisheries is a state subject. Each state has its own fisheries department, either under a separate ministry or under some other ministry – Animal Husbandry and Dairying, Agriculture, Forestry and Conservation, etc. In addition to fisheries agencies at state level, there is a Department of Fisheries in the (central) Ministry of Agriculture and Cooperation, under a Joint Secretary of Fisheries and the Commissioner of Fisheries Development. There are several deputy commissioners, one for each sector, who provide guidance to state governments in the implementation of fisheries development programmes. Aquaculture is the concern of the Deputy Commissioner, Inland Fisheries.

In the *Philippines* certain responsibilities relating to coastal aquaculture, for example, the issuing of licences to farm the foreshore and shallow waters, have been delegated to local municipalities.

Responsibilities for emergency intervention

Many countries have found it necessary to develop policies and use policy instruments in reaction to urgent situations which emerged unanticipated in their aquaculture sectors. For example, countries interested in developing coastal aquaculture have had more than their share of problem issues. In addition to compliance with the usual health controls for marine food-products, there has been a need to protect production sites which competed in areas where there was a long tradition of harvesting 'common property' aquatic resources from the wild. Coastal marine laws which existed typically related to navigation and rights of passage, and the use of the foreshores and seabed, but modern marine culture (and in some cases culture in freshwater lakes and brackish-water lagoons) required usufruct rights (the rights of all to use and enjoy resources without impairing their substance) over the column of water in which the species were being grown. Consequently governments have suddenly had to take into account the interests of competing sectors (such as tourism), conservation issues, and environmental protection.

Where market opportunities and technical capabilities in the private sector already existed, then, organizationally, governments have only been called on to provide minimal management inputs to regulate orderly development. Provided markets were available, private enterprise has been able to expand production to meet demand at a rate commensurate with market capacity and economic return. In some cases, where demand has far exceeded supply (for example, when the investment in marine shrimp production caused widespread destruction of coastal mangrove areas in countries of Asia and Central America, and the over-harvest of juvenile marine shrimps), governments were called upon to intervene to prevent malpractice occurring, such as over-exploitation of resources from a limited area, inadequate safeguards in respect of product quality, environmental and pollution control, and monopoly practices and price manipulation. Conversely, governments have also intervened when a large number of constraints were restricting the potential of the sector, and when accelerated aquaculture production was required to compensate for an anticipated decline in natural harvests.

Government intervention can be used when too many constraints are restricting the sector's potential

'Although aquaculture production in Thailand is unlikely to increase greatly in the short term, in the longer term it has the potential of increasing fish production to meet local consumption, and could make a major contribution to export earnings. Development of aquaculture is constrained by some of the same factors affecting marine capture fisheries, including pollution, land and water-use conflicts in the coastal zone, and inadequate product handling. Additional constraints include a lack of good quality seed to stock ponds as private sector shrimp hatcheries are not controlled, and production from government hatcheries is low. Much of the technology needed for significant advances in production is well known, but is not being transferred to the private sector in a timely manner. Government aquaculture staff lack commercial experience to train small farmers so that the country can better capitalize on its aquaculture resources and technology base.'

Source: Asian Development Bank, 1985.

The example of Thailand is a classical case for government intervention; firstly, on the basis of a limited time-frame to upgrade hatchery production to meet present demands, and to organize the transfer of technical knowledge through training to ensure that ongoing demands are met through improved and increased private hatchery production.

Serious commitment to the sector

In many developing countries, where technical capabilities have been low or virtually non-existent, and where few if any market potentialities were apparent, the public sector has not been as committed as necessary to develop any useful aquaculture sector. Strong rhetorical encouragement in policy statements, particularly by countries in Africa, for fish culture development have not been backed by correspondingly strong support services. The results have therefore been counterproductive, and set back aquaculture development in the region. The combined production figures of the 36 countries of Sub-Sahara Africa have shown little growth over a decade, and many individual countries have shown a distinct decrease in national production.

Without firm government commitments to support aquaculture development actively, there is little likelihood of success under present circumstances

Two studies reflect the contrasting commitments in Sub-Sahara Africa (where there is virtually no sustainable aquaculture) and in Asia (with prodigious annual production):

(a) In Africa, south of the Sahara, it was found that 'government support to the sector has been little at best' and that 'only a few countries in the region have identified aquaculture as a priority *vis-à-vis* agriculture, and actively support the development of the sector'.

By comparison:

(b) 'Formal policy statements and plans reveal that most countries in East Asia have gradually shifted emphasis and support from the traditional marine capture fisheries to inland fisheries and aquaculture.'

Sources: (a) ADCP/REP/89/36, and (b) ADCP/REP/88/31, FAO, Rome.

Develop an infrastructure

Due to the relative immaturity of the aquaculture sector in most countries, both developed and developing alike, and the interaction of the objectives noted above, infrastructure support presently given to the sector by most countries is still only evolutionary. Consequently, it is often liable to marked and sudden changes in the light of greater experience and altering circumstances. Moreover, in some developing countries which have achieved independence from colonial rule, with the passage of time and exposure to new ideas, they have deliberately replaced inherited cultural patterns with something different, and not necessarily more appropriate to their new needs.

New infrastructure support may deliberately set out to be different from past patterns

In anglophone countries of Africa and Asia there was usually an established corps of fisheries personnel within the Civil Service. New aquaculture officers and extension agents have now been assimilated into this corps. In most francophone countries, with the exceptions of Senegal and the Cameroon, there is no longer a special corps of fisheries staff.

The immaturity of organization and management experience in many countries has frequently led to a confusion of interests, conflict of responsibilities, duplication of effort, and a lack of clear direction of resources (already in short supply) to the points of greatest impact. This has been especially evident in developing countries where lack of training and experience has shown more tangibly and to greater effect.

Apart from the production process itself, which in all but centrally-planned economies is usually left to the private sector, aquaculture, in common with other primary production sectors, depends upon a range of organization and managerial support services for its development and regulation. Some of these services can only be supplied by government, others can arise from within the industry itself, and some can be provided by educational and research institutions, such as universities and technical training colleges. All these support services ultimately aim at providing a coherent and comprehensive 'whole', which will enable the industry to stand on its own feet and pay its own way within the community and country.

No panacea is available for general application to effect successful development and administration of aquaculture. Each country has evaluated its own requirements in the sector differently, and established an infrastructure support system appropriate to its own needs and political, social, and economic policies.

One final point is the influence which national and regional policies have had on the stratification of structures. For example, at a national level, aquaculture in Canada may be relatively unimportant and command little in the way of a national organization and management structure, but at a regional level, in the coastal provinces, it plays a crucial role in economic and social development and commands a relatively sophisticated infrastructure. Similarly, while the 'fisheries' ministry has had its own organization and management structure(s) for directing production into the food-supply chain to the consumer, another agency of national or regional government may have become engaged in aquaculture for very different reasons. For example, in Pakistan, the Federal Water and Power Development Agency has an aquaculture arm which takes in regulating all large water masses, but uses certain cultured fish species to control vegetation. In this case, production of fish is incidental to other objectives but has not disrupted market forces or affected the objectives of the food production agency.

5.1.3 Aquaculture infrastructure in the hierarchy

The central administration

One of the issues which has retarded countries from translating policy ideals for aquaculture into effective realities has been the quandary of where to place administration responsibilities for aquaculture within the government infrastructure. A summary of the variable selection options carried out by one region alone, namely Africa, demonstrates the problem (see Figure 5.4). In several cases, the delegated ministries are multifaceted agencies. This has magnified the problem of translating policy intentions into practical development support and clear decision-taking. Not surprisingly, aquaculture has been greatly neglected if the prime responsibility of the agency is another key economic sector, such as land, tourism, wildlife, or natural resources, etc., which earn foreign exchange.

The quandary of governments for placing the aquaculture administration most effectively into the national framework has often been caused by the sector proceeding in advance of any recognition by government, and the fact that it has required particular support and regulation to enable it to expand in a healthy and progressive manner. Consequently, few govern-

Angola	Agriculture
Burkina Faso	Environment et du Tourisme
Cameroon	Livestock, Fisheries and Animal Husbandry
CAR	Eaux et Forêts
Chad	Tourisme, Artisanal, Eaux, Forêts et Chasse
Côte d'Ivoire	Agriculture et des Ressources Animales
Ethiopia	Agriculture
Gabon	Eaux et Forêts
Guinea Bissau	Dévéloppment et de la Pêche
Kenya	Tourism and Wildlife, Regional Development, Science and Technology
Lesotho	Agriculture and Marketing
Malawi	Forestry and Natural Resources
Mauritania	Pêche et de l'Economie
Niger	Agriculture et de l'Environnement
Nigeria	Agriculture, Water and Rural Development
Sudan	Agriculture and Natural Resources, Animal Resources
Swaziland	Agriculture and Cooperatives
Tanzania	Land, Natural Resources and Tourism
Togo	Developpement Rural
Uganda	Animal Industries and Fisheries
Zimbabwe	Agriculture, Lands Resettlement and Rural Development, Natural Resources and Water Development

Figure 5.4 The position of aquaculture administrations within the government infrastructure in Africa (Source: ADCP/REP/89/36, FAO, Rome).

ments of developing countries have been able to react decisively in organization and management terms, and particularly when faced with demands from differing ministries which may have had an interest in the sector.

Some governments have been very flexible toward the delegation of authority for aquaculture, according to where such development is taking place physically, and have involved several separate ministries. Within each of the respective authorities, these countries have built up the necessary expertise and capabilities to exercise their functions.

Several governments have been flexible in delegating responsibilities for growth of the sector

Under the regimen in the USA, freshwater pond culture has been the responsibility of the Department of Agriculture, as decisions were necessary in terms of land-use alternatives, and the integration with animal and poultry production in terms of extension. Responsibilities for salmon production and sports fishing have remained with the Department of the Interior (Fish & Wildlife Service). Aquaculture production in the coastal zone and offshore (excluding migratory salmon enhancement), has been the responsibility of the Department of Commerce, which handles all marine fisheries. Matters concerning the environment were the responsibility of the Environmental Protection Agency.

Broad delegation has been made most typically with countries with extensive areas of land and water, such as USA, Canada, Brazil, India, Indonesia, and Nigeria, including the advantages of further responsibilities being given to the individual states as well as agencies targeted to develop specific resources or regions. In time, these governments might reduce their involvement (other than in a regulatory context), and hand over responsibilities to the private sector, through a federation of producer associations, to enable their resources to be diverted to new development concepts.

In most countries the ministry of agriculture, or its equivalent, has been responsible for administration of aquaculture. This has usually been designated as a sub-division of the ministry's fisheries directorate or section. Despite the expanding scale of aquaculture production in many countries, and its potential for increasing food supplies and contributing to economic growth, aquaculture generally has occupied a relatively minor position within the administrative structure of government (see Figure 5.5).

Significant in the example (Figure 5.5), and typical of the administrative structures in many other countries, is that, even within the fisheries directorate, aquaculture has not merited a full-time assistant director's time, but is grouped together with other and disparate responsibilities. It also appears

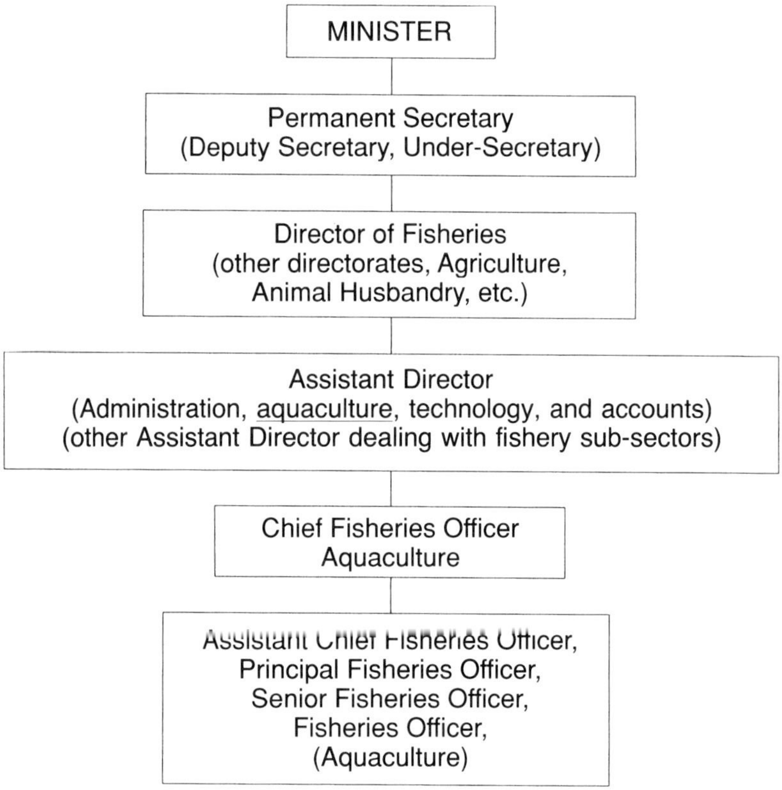

Figure 5.5 Simplified organization structure of aquaculture administration within the Ministry of Agriculture, Nigeria.

unlikely that the assistant director requires any training or technical knowledge of aquaculture. With this structure, it is highly probable that aquaculture has a hard task to hold its position within the fisheries directorate in the competition for recognition and funds, and making its mark within the overall ministry as a vital new field. However, this might be overcome by any influential personalities among the senior staff who are not concerned with a strict interpretation of the organizational structure as laid down.

Other institutions within the hierarchy

Governments of most developed countries and several developing countries of Asia have recognized their responsibility of providing institutional support for the necessary services. Many of these services have been provided by typical government institutions which have other primary functions (such as agriculture research, or the agriculture extension service). Some governments, however, have viewed aquaculture as sufficiently distinct, or of special economic or social importance, to justify establishing a specialized research, education, or training institution, or separate extension services. Aquaculture research has been carried out both within universities and at research stations, funded either by the government or by a provincial authority. Funding has been both total, for a specific laboratory to undertake a major programme, or targeted at specific projects, or a combination of the two. For the most part, university research has been directed to fundamental science, leaving government centres to resolve problems of more immediate and practical importance. On the whole, research activities have been well decentralized throughout every country.

Where aquaculture has been accorded high priority for specific economic reasons, such as in Japan, Israel, and Taiwan, governments have established a coordinated research agency charged with responsibility for all aspects of aquaculture research. In Japan, for example, the agency is the National Research Institute of Aquaculture. In addition to research, the agency has other specialized functions, such as education and training support. Extension services to the industry have also been provided by government through universities, or by a specialized agency, or by a combination of all or some of these organizations.

In the late centrally-planned economies (for example, in Hungary and Czechoslovakia) state farms, hatcheries, marketing organizations, and other institutions have been designed specifically for the development and support of the sector (see Figure 5.6).

Cuba established a relatively unique and specialized organization called the National Aquaculture Enterprise (see Figure 5.7). This is a state-controlled organization set up to research and develop aquaculture in the country, and to establish and run production, processing, and marketing operations. In other words, the institution was mandated with comprehensive and integrated functions to develop and manage aquaculture development in the country.

The National Aquaculture Enterprise is an unusual example of governmental support for aquaculture development. It was established under the control of the Ministry of Agriculture but is a relatively independent organization, i.e. a parastatal. It is also unique, however, in that its range of functions is very wide and comprehensive, covering support and production functions which are normally divided between government departments, institutions, and the production sectors. The Enterprise is said to be self-financing with income from sales re-invested in development programmes.

Although the National Aquaculture Enterprise of Cuba could well be considered a parastatal organization, there are obvious doubts as to the degree of autonomy it enjoys, in legal terms, and also the degree of self-financing.

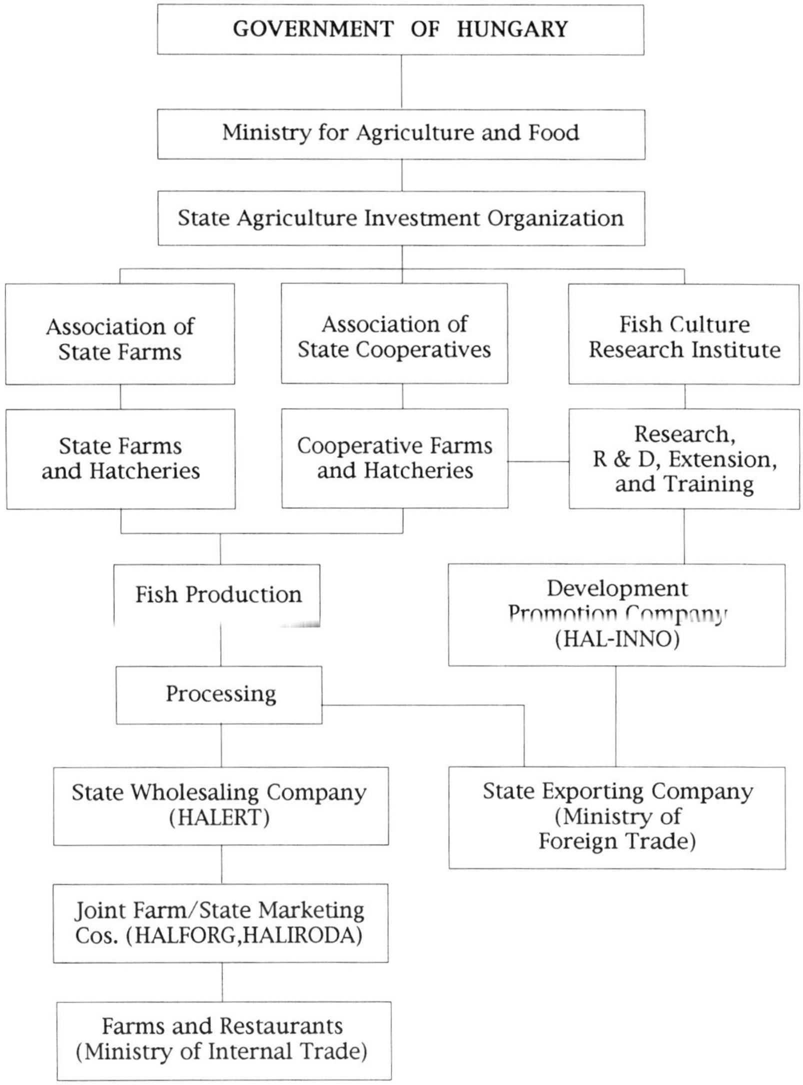

Figure 5.6 Simplified organization for the aquaculture/fisheries industry in Hungary.

Parastatal organizations

Many governments have provided state support through the creation of parastatal organizations or authorities. These organizations have been given broad terms of reference and delegated authority, or have been specifically directed toward a principal function with various ancillary responsibilities.

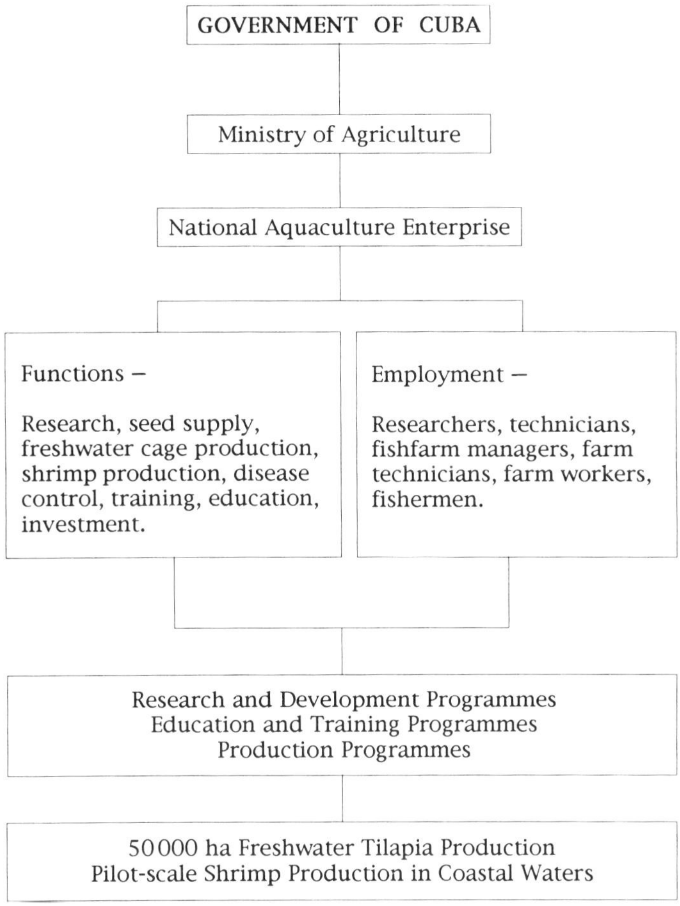

Figure 5.7 Institutional support for aquaculture in Cuba: the organization of the National Aquaculture Enterprise Agency.

Responsibilities characteristic of parastatal organizations with broad objectives

Objectives and functions of the New Zealand Fishing Industry Board include:

- promote the fishing industry in New Zealand,
- promote means of expanding the fishing industry and ensure full use of resources,
- promote sales of fish and fish products for consumption in New Zealand,
- ensure adequate supplies of fish and fish products at reasonable prices,

- promote means of ensuring proper standards of quality and hygiene in respect of fish and fish products,
- coordinate domestic and export marketing of fish and fish products,
- license exporters of fish and fish products,
- resolve problems relating to economic production of fish and fish products,
- promote a greater degree of coordination within the fishing industry,
- advise government and industry of development of the industry, and
- promote means of obtaining finance for development of the industry.

It is immediately apparent that the New Zealand Fishing Industry Board had no particular responsibilities for or toward the aquaculture sector as such. Its terms of reference, however, were broad enough to permit its intervention to the benefit of the sector, if it was persuaded to do so.

The mandates of the Fish Marketing Organization of Thailand, and the Fish Marketing Organization of Hong Kong were more direct in identifying their responsibilities.

Responsibilities of parastatal organizations with directed objectives

Objectives and functions of the Fish Marketing Organization of Thailand include:

- develop wholesale fish markets, local fish markets, and the fishing industry generally,
- arrange, control, supervise, and render services to fish agents,
- improve the living standards and promote the welfare of fishermen and their communities, and
- encourage and support the establishment of fishermen's cooperatives.

The objectives of these parastatals were to be achieved by provision of fish markets and landing facilities, promotion of fisheries, and operation of fisheries development projects. Although establishment of the parastatal was originally in the context of developing and assisting the marine capture fishery, there was nothing in the terms of reference to preclude the aquaculture sector, and it did not.

In Chile two quasi-government, or parastatal organizations have provided considerable support to its rapidly growing aquaculture sector. These are ProChile, which markets national production, including aquaculture products, and Fundación Chile, which supports research and development in communications and resources, and hence includes aquaculture in this purview.

As yet, there are no parastatals concerned only with aquaculture. This is in keeping with the fact that no country has, as yet, established a national department of aquaculture independent of associated agriculture and/or marine fisheries departments, and their responsibilities.

The record of parastatals is highly variable. Many parastatals, particularly in the developing economies, have failed to achieve the aims and objectives set out for them. This has been due to a variety of causes, such as management inexperienced in the conditions of the industry, over-ambitious or ill-defined organizational objectives, interference by government in day-to-day management, poor infrastructure in other relevant sectors (such as transportation), and organizational activities which are inappropriate to the industry's problems and constraints.

Parastatal organizations are a feature of both centrally-planned and free-market economies. In centrally-planned economies they have a degree of autonomy from government departments, which enables them to perform many of the market-related functions normally undertaken by the private sector in free-market economies. The two benefits of a high degree of autonomy and being financially self-sustaining have been the key to their effectiveness, particularly in several parastatal organizations in Hungary which operate in independent but integrated systems.

5.2 The private sector

5.2.1 The principal role of the private sector

Governments have accepted a varying role in the support of the aquaculture sector. With certain exceptions, this role has excluded direct involvement in production operations, which is the principal role of the private sector on which the industry depends. The exceptions, which include the enhancement of natural fisheries (see section 1.3.2) and the direct involvement in state-farm production, whilst endeavouring to follow commercial principles and practices, are usually targeted in support of national production.

In general the majority of private enterprises are small-scale and, as previously observed (see section 1.3.1), aquatic farming is predominantly an individual entrepreneurial operation for the production of fish and shellfish for sale on local markets. Most production enterprises are usually owner-operated, or based on some form of tenancy agreement with either a (corporate or individual) landlord, or with government (in some form or other). Tenancy or leasehold from landlords has occurred where lands are already fully owned, such as in the great estates of Scotland or some of the coastal lands of Indonesia and the Philippines. Leasehold from government has occurred on new or marginal lands owned or repossessed in the national interest to expand primary production to overcome food shortfalls (present or anticipated), or to stimulate exports in selected products. Nonetheless, the basis for the majority of aquaculture production has been placed in the private sector for its realization.

In its simplest form, aquaculture, as in many poor countries of Africa, has been an alternative form of subsistence farming. At the other extreme, as in most of the developed countries, it has become a highly sophisticated system of vertically integrated company operations, from high technology production, through processing, and marketing direct to the consumer. Between these extremes there has been developed a range of intermediate organizations and managerial systems world-wide (see section 1.3).

The significance of this range of private sector investments is that its dependence upon government-sponsored support services has been inversely proportional to its degree of sophistication. In other words, the less sophisticated the practice, the greater the dependence on government support. The application of government regulations and controls in respect of land and water use, quality, health, and hygiene, etc., has remained true in respect of all operations, regardless of size and nature. Therefore, most governments have had to evaluate the status and condition of the private sector within the national aquaculture sector before determining the nature and scope of their support services.

This has not been easy for many governments, which have consequently encouraged the private sector to start their own 'support services' through the traditional cooperatives or producer associations, covering a range of activities. Being close to the producers, these services have often been much more effective than those provided by central or provincial governments, or even universities. Governments have therefore confined their support to initial training and refresher courses, which have been financially less expensive than a commitment to a large and expensive public extension service.

5.2.2 Secondary roles of the private sector

Organization within the industry

Structured private sector organizations, such as federations or associations of producers, and marketing or trade associations, have played an important part in the successful development of many sectors. This has been most productive where the aquaculture organization has been grafted onto an established and successful organizational system of producer associations already in place to service other sectors of the primary production industry, say, agricultural or fisheries cooperatives. Unfortunately, in practice, cooperatives have not been as generally successful as anticipated. This applies on a world-wide basis throughout the developing countries.

Cooperatives have not been very effective as an organization of the private sector to increase development

The main reasons for failure of producer cooperatives include:

- poor motivation of members, partly due to the fact that, historically, cooperatives have been the creation of government rather than arising from the demands of members,
- poor management, frequently leading to financial loss and subsequent loss of trust by members (usually due to inadequate training and pay-levels for employees),
- strong traditional links between fishermen/fishfarmers and 'middlemen' who provide regular, reliable supplies, credit and markets.

As a result of past failures many developing countries have had to embark upon retraining programmes to rectify known deficiencies, and strengthen national cooperative organizations without placing too much reliance on their potential for success. Where cooperatives have been established as a result of members' demands, however, or where fishfarmers have got together to establish their own producer associations, the chances of success have been considerably higher.

Caution is required in placing reliance on cooperatives to develop the sector

'It is recommended that the Government [Indonesia] adopt a cautious approach to the expansion of fisheries cooperatives. Management and operation of existing cooperatives should be strengthened before new cooperatives are formed, and Government should not initiate cooperatives as a means of competing with the private sector.'
Source: Asian Development Bank, 1983.

It is interesting to note the placement of cooperatives outside the private sector. This reflects an understandable but common mistake of looking upon cooperatives as an 'arm of government', when they are an expression of one form of united private sector activity. This misconception has probably contributed to the underlying failure of cooperatives in developing countries, (where they continue to be thought of in quasi-government terms) compared with their place in developed countries (where they more nearly approach successful businesses in conventional company form). Also, in many developing countries, cooperatives are seen and used as government policy instruments. Such misconceptions more or less have condemned them to failure as commercial and economic entities, and have often led to their complete restructuring with every change of government.

Given the relative efficiency of private sector marketing, cooperatives have had to be self-supporting to play an effective role in development of aquaculture. A fisheries sector study in Thailand, by the Asian Development Bank in 1985, recognized the paramount importance of improvements in management as a prerequisite to meeting government objectives for the fisheries sector. It was noted that the fisheries resource was not being managed effectively, that environmental degradation was a serious problem, and that there was little effective management of the coastal zone. Institutional strengthening in these three areas was seen as the key to effective resource management. But it was also stressed that rapid development in the fisheries sector, and the achievement of production targets, would be heavily dependent on activities in the private sector. Whilst the main thrust of the study was in relation to the marine fishery, the principles propounded applied equally to aquaculture, and it is of interest to note that progress has been made in the recent establishment of aquaculture (land settlement) cooperatives which are generating high profits.

Financial independence

A major problem which has faced most small-scale farmers, particularly in developing countries, has been their limited ability to acquire the necessary capital to break out of the debt cycle tying them to local money-lenders. Once the break has been made, each enterprise has become a viable commercial proposition, and the acquisition of further finance has become progressively easier as private sector financial agencies show increasing interest in providing monetary support. In the long term, successful development of the sector has led to competitive investment offers more favourable to each entrepreneur, and finally financial independence.

Government investment and support has been most beneficial in the initial stages of development when financial problems have beset the potential farmer. These have been the high costs of capital works, such as the construction of ponds and the lease of land, and supplies for start-up operations (seed, feed, and fertilizer). Considerable government support has been provided by nominal rates for the land (for a period of years before commercial rents are then levied), and capital grants and/or loans toward the costs of capital works, construction of ponds, cages, or a small hatchery. Support has also been in the form of tax exemptions or deferred payments over a period of years for private sector investors prepared to undertake such capital works, or even provide working capital where necessary. The permutations and commutations used to stimulate private sector investment are manifold, and are discussed in detail in Chapter 7. For the most part, aquaculture has only developed in a secure and stable manner when the private sector has been persuaded to invest, and at a significant level. Where development has remained tied to government funding its growth has been stultified, and lacking the ability to branch out of its own accord into new technologies and managerial systems.

Processing and marketing

In countries with a free market economy, the private sector has played a major role (and probably *the* major role) in processing and marketing, once opportunities developed beyond production for the local market. The private sector was then able to influence market forces by the introduction of alternative product forms and products at competitive prices.

Customer preference, which decides the acceptability of aquaculture products, has been dependent on a large number of external variables not always easy to quantify and evaluate. Once aquaculture producers became dependent upon marketing and processing services to sell their products, they placed themselves in the hands of professional processors, distributors, and marketers who had the expertise and experience of transferring raw products to the ultimate consumers in the most acceptable form and at acceptable prices. Such services have had to be paid for, regardless of who provided them. The private sector is pre-eminent in this sphere, and any direct government intervention has been generally costly and inefficient. Typical interventions have been government price controls, either by squeezing the producer or subsidizing costs out of general revenues, or by limiting the availability of alternative foods, thus forcing the consumer to pay exorbitant prices. For these reasons governments have often opted to use parastatals within the marketing system as they have a marketing monopoly.

Organization of supplies

Few producers have had access to their own sources of feed and seed, such as the marine fishfarmers in Hong Kong. The vast majority have had to depend heavily on other sub-branches of the private sector, such as hatchery enterprises and feed manufacturers. The preparation of pelleted feed supplies has readily been undertaken as part of comparable production for livestock feeds generally, but usually much more competitive in price due to the specialization and relatively small quantities involved. Although it is usual for government hatcheries to initiate seed production, it is normally well beyond their capacity to attempt to supply seed to all producers. The policy has been to encourage cooperative hatcheries to meet the wider demand, but mostly the private sector has taken over seed production, particularly for high-value species.

Other inputs of working capital, such as cages, nets, floats, pontoons, pumps, etc., have all been supplied by the private sector, and often under credit terms which represents financial investment in the sector at a grass-roots level. Frequently, the private sector has provided these inputs with supporting extension and training services. Clearly much of this assistance has been with the self-interest of the suppliers, and consequently these services have not entirely replaced the unbiased extension services which government or other organizations, such as universities and research establishments, have offered to producers.

5.3 The practical issues of organization and management

The final section in this chapter addresses the issues involved in translating aims and objectives into practical realities.

The organizational framework

A great deal of administrative skill and foresight is required to interpret intentions into management through an appropriate organizational structure. The options are:

- creating a new structure, or
- adapting an inherited structure originally designed to achieve differing objectives.

The former is usually the easier of the two, but the latter is probably the reality and state of affairs facing a senior administrator. Dovetailing new needs of aquaculture into established practices (of capture fisheries or agriculture) is not an easy task.

Whilst recognizing that, once started, aquaculture will continue to be practised with or without the help of government, the organization and management of the sector is largely dependent on it. The degree of interest shown by government, its policies, and bundle of policy instruments, all determine the future nature, structure, and health of the industry.

There are no hard and fast blueprints which guarantee the creation of a successful aquaculture industry by government design alone. External factors outside the control of government all play their part, even though it may be possible for government intervention to ameliorate or modify their impact; for example, the availability (and value) of export markets for marine shrimp and Atlantic salmon which have changed as more countries became competitive producers.

Internal factors within the industry are conducive to success above all else. These factors are theoretically within the control or influence of government. The need is for a discreet blend of encouragement combined with imposition of legislative measures necessary to ensure self-discipline within the sector. 'Rogue' activities by a few irresponsible operators at any level can prejudice the welfare of all, and this must be widely appreciated in the industry.

In order to create the right balance of limited control and imposing self-discipline, it is imperative for government to maintain an open and effective information service, which can carry the industry opinion with it, especially in time of crisis. Although the private sector carries the industry in operational and financial terms, it always looks to government for guidance and assistance when the latter is called for. On the other hand, governments expect the industry to sustain itself, and not depend on grants and subsidies.

In organizing the sector with a view to development, the first action to be taken by government is to evaluate its current status, in socio-economic terms, and attempt to assess its long-term potential given requisite encouragement and support. At this stage it is possible to identify management deficiencies which must be corrected if the sector is to grow satisfactorily. Institutional provision can then be built into the structure; for example, creation of credit arrangements for low-interest loans to first-time farmers, or the introduction of 'codes of practice' for processors handling export products.

The evaluation may also attempt to forecast some optional alternatives regarding the direction and pace of future development. The ability of the government to develop within a flexible framework, to take advantage of opportunities as they arise or to offset unforeseen delays in development, helps to sustain confidence in the sector and to ensure stability. It also indicates that the government is leading from the front and not merely, *post facto*, responding to events as they occur.

Selecting appropriate policy instruments

In principle, there is a wide range of policy instruments available to governments seeking to plan and manage sector development. The real availability and appropriateness of particular instruments are, however, determined to a substantial extent by a number of factors which influence the ability of governments to *apply* particular instruments or measures. It is important, therefore, to take account of the general political, economic, and administrative contexts in which planning and development efforts are being applied.

First, the question of what degree of government intervention is appropriate or desirable is an issue around which different governments are deeply divided, and which has given rise to fundamentally different approaches to economic development. This difference is most acute, for example, between centrally planned economies and capitalist or free-market economies. In the former there is a high (and often total) degree of state control of the processes of production and distribution, and in the latter control is less direct. Therefore, clearly, the use of policy instruments for the development or management of the aquaculture sector is to a large extent a matter of the type of economy within which sector management is taking place; and what is appropriate in one context is not necessarily so in another. Furthermore, governments have differing views on the general desirability or effectiveness of particular instruments. For example, fiscal subsidies may be totally precluded or widely utilized for developing the aquaculture sector, depending on overall government policy.

Second, the appropriateness of a particular instrument to aquaculture development within any country is a matter of whether the administrative or institutional capability exists within the country to apply it effectively. For example, many governments do not have the level of administration or institutional infrastructure, particularly in remote rural areas, required to run sophisticated business support or training schemes which some forms of aquaculture development may require. Thus the choice of any policy instrument or instruments is greatly influenced by the capabilities of governments.

Third, the suitability of particular policy instruments is also influenced by the policy objectives of governments. For example, some countries may place high priority on the production of cheap fish for local consumption, and others on the production of high-value fish for export. The appropriate policy instruments for financial support or incentives are therefore very different in each case. Countries also differ in the relative level of priority given to development of aquaculture itself, or to its competitors for resources (such as agriculture or tourism), or simply to conservation of the environment, and these different objectives also give rise to differences in the policy instruments applied.

Fourth, the type of aquaculture industry and its current state of development are also important determinants in the choice of instrument. The needs of industries dominated by high-technology intensive production systems are very different from those made up of low-technology extensive systems. The appropriate policy instruments, therefore, are equally different. For example, policy instruments to encourage investment in intensive systems producing high-value commodities include such things as subsidies on feeds, possibly state feed suppliers, or grants for capital equipment. None of these is necessarily appropriate to encourage extensive farming, which requires few inputs except credit for pond construction and possibly seed from a state hatchery.

Finally, countries with well-established sectors have different needs from policy instruments from those where sector development has yet to take place. For example, if the majority of producers are relatively new to the field their need is for government financial assistance, in the form of soft loans and grants, until their production systems prove to be viable and attractive to commercial sources of finance. Developed sectors have proven technology and viability and attract finance from commercial sources, and consequently government instruments are more concerned with duties, taxes, and quotas, rather than loans and grants.

In summary, there are a number of basic factors affecting the choice or appropriateness of policy instruments. The political and economic system within which the particular industry is located is obviously an important determinant for the use of policy instruments, as there are major differences in fundamental philosophy and approach to development between centrally planned and capitalist economies. The capability of the government and national institutions to apply policy instruments effectively is also a factor which determines the choice of instrument or instruments appropriate to any particular case. Countries with highly developed and experienced public services and infrastructure are generally better able to apply a wider range of policy instruments than those where such services and infrastructure are limited or underdeveloped. Moreover, the government's goals for the industry affect the choice of instrument, as does the existing state of development which aquaculture has already attained in the country.

For whatever purpose policy instruments are intended, it is important to note that their use in the management of the aquaculture sector is a complex matter which is not open to easy generalization. The choice of any instrument or instruments must be made within the context of all the factors summarized above, which are peculiar for each and every country. Consequently there are no common models which may be readily applied.

The degree of commitment, as indicated by government policy statements and choice of policy instruments, does not always correlate closely with current levels of national production. To a considerable extent this is due to the largely traditional nature of aquaculture production. In many countries where aquaculture production is high (predominantly in Southeast Asia), their industries have evolved over a long period of time without the assistance of government policy or policy instruments. The long tradition of aquaculture in China, for example, which dates back more than a thousand years, has resulted in extraordinarily high annual production levels (over 6 million t) without use of modern policy instruments. Conversely, Canada, which has in recent years demonstrated considerable commitment to organization and management, and the creation of special policy instruments for the development and control of the sector, is as yet far from having an important sector (about 35 000 tonnes per annum).

Therefore the use of policy instruments in a formal manner by governments has not been a *sine qua non* of the aquaculture development process. Rather, planning and the application or special use of policy instruments for aquaculture, are very recent developments. Moreover, plans are mostly being prepared and applied in the pursuit of a range of objectives which are geared to meet short-term needs in the different countries, rather than long-term perspectives for national investment in all its forms.

In the following chapters (6, 7, and 8) the main types of policy instruments available to government administrations for the organization and management of the sector are examined and, where possible, examples of their use are given. Their appropriateness varies according to the economic system, government administrative capability, policy aims for the sector, and type and level of industrial development. The selection of illustrations, to a large extent, reflects the diversity of situations in which the instruments may be applied.

Implementation of policy instruments

Having assessed the status and potential of the industry, and determined the policies and policy instruments required to develop the sector effectively, the next step is to implement them. Implementation is rarely, if ever, achieved by one simple act. Typically it is a step-by-step approach over a period of time. This is because, among other considerations, it takes time to train staff to undertake their new duties effectively, otherwise the impact of the policy is lost and the effort wasteful of money and staff resources. The lack of training and implementation expenses have been primary reasons for the failure of many government policies for aquaculture, particularly in developing countries anxious to catch up in the general race to establish an industry.

Training is an integral part of development when establishing a new sector

A review of the aquaculture sector in Sub-Sahara Africa found: 'Departments tend to be heavily staffed. The staff complement in the Inland Fisheries Division in Ghana in April 1988 was six fisheries officers, 111 technical officers, and 127 technical assistants; but most of the mid-level personnel are not trained and there are no specialist staff'. Ghana's inland fisheries production annually is about 40 000 tonnes, and aquaculture 450 tonnes.
Source: ADCP/REP/89/36, FAO, Rome.

Implementation becomes another organization and management problem when there are several interested agencies of government, each of which considers it has a part to play in sector development. Although this in itself is no bad thing, as there is much scope for assistance in any emerging industry, difficulties arise when there are no clear mandates and responsibilities identified for each by the government, or a structure for coordinating the whole pattern of support into a coherent programme.

Many agencies of government consider they have a part to play in developing the sector

'In Paraguay aquaculture falls under the Dirección de Caza, Pesca y Piscicultura of the Ministry of Agriculture and Livestock, but at the lowest hierarchical level. In spite of the organic law that designates this Directorate as responsible for aquaculture, the lack of human and economic resources impede the execution of its functions. In practice, the Fish Farming Division of the Faculty of Veterinary Sciences at the national University of Asunción is the main promoter of aquaculture throughout the country, having as its base of operations an experimental farm which includes 4 ha of fish ponds.'
Source: ADCP/REP/89/39, FAO, Rome.

However, as in the case of Paraguay, the activities of other agencies are usually tolerated until there is a question of 'regulating' the industry. Then the need for staff and funding provisions becomes serious.

It may be assumed that inadequate funding for both staff and functions in organizing and managing the sector is more widespread than is admitted, and that many of the organizational charts and management systems outlined are yet to be attained in effective operational terms. Having said this, the care and attention devoted by many countries, both developing as well as developed, to the preparation of organizational and management systems suitable for local needs is impressive, also demonstrating where and how parastatals, other departments, and universities contribute to the whole.

At the risk of apparent over-simplification, it is suggested that development of the aquaculture sector in organization and management trends (or bias), insofar as government inputs are concerned, moves from 'promotion' through 'monitoring' to 'regulatory' phases. Obviously these are not clear and distinct phases, and at all times there will be an element of the other two inputs within the current management system. Differing governments will apply their managerial inputs in differing ways. For example, some governments may decide to place regulatory functions exclusively within the remit of one department or agency, whilst promotional efforts are applied via another agency.

Governments apply managerial functions differently

'In El Salvador the Centro de Desarrollo Pesquero (CENDEPESCA) is responsible for the aquaculture sector; but its functions are exclusively regulatory. All operations to promote the sector are under the control of Agricultural Regional Authorities. This institutional structure prevents CENDEPESCA from having direct authority over extension and promotional personnel; and thus places a distinct constraint on the mechanism for the support and control of the sector.'
Source: ADCP/REP/89/39, FAO, Rome.

It is more usual for a government to make one central agency responsible for both the promotion and control of the sector, so that any change from one emphasis to the next is introduced gradually and appropriately. Changes of this nature are typically introduced at different times, and possibly at a different rate in different regions. Management flexibility in these cases is a valuable asset, and should be built into the organization infrastructure. The fact that one agency is nominated to promote and control the sector should not inhibit other agencies from playing supportive roles provided that they follow the mainstream policies and plans of the lead agency.

In the promotional phase of establishing the aquaculture sector, the main concerns and objectives of the government are to get the sector functional, effective, and producing. Whether to supply local demand or export markets is immaterial in this context. However, possible negative side-effects arising from an emerging industry are often ignored. Recognition that there may be negative effects comes more slowly, and notice usually comes from other agencies of government; for example, environmental and conservation agencies which may see detrimental impacts from new industry. Regulation is then introduced for acceptable operational criteria, and a more mature and stable industry results.

Motivation

One of the most important aspects of management of the sector is the human resource. This applies not only to the public sector but also to the parastatal and private sectors. In the private sector, enlightened self-interest usually provides sufficient stimulus for those concerned to give their best in the interests of the industry. In the public sector, and the quasi-public parastatals, motivation has to provide the necessary stimulus.

Motivation is difficult to define, and is often harder to stimulate. It is promoted, however, by financial and promotional incentives, certain privileges, and good communications. It is frequently stifled when any or all of these factors are absent, or grudgingly conceded. Given well-motivated staff, almost any organization can be made to function reasonably well, provided organizational deficiencies are not basically flawed.

Well-motivated staff can make any organization work well

'In Panama, since 1976, the National Aquaculture Directorate (DINAAC) of the Agriculture Development Ministry has been in charge of the implementation of the country's policies toward the aquaculture sector as dictated by the National Development Plan. The lack of a National Aquaculture Programme has not impeded DINAAC from conducting an active scheme for the development of the sector at all levels. Through a sound but flexible internal organization, DINAAC maintains an active role in extension, research, promotion, regulation, and control in both the rural and industrial sections.'
Source: ADCP/REP/89/39, FAO, Rome.

In conclusion, there is no simple formula or organization and management chart which can be recommended as ideal for successful development of any individual aquaculture sector. This is because of the wide range of differences between countries, particularly their geographic, economic, social, and even historical circumstances. Each country has to evaluate its own needs and seek its own solutions. However, no organization and management structure should be looked upon as permanent, particularly as growth and economic importance can change rapidly in less than a decade (see Figure 5.8). There will be continual pressures upon the structure which will require changes to be made to meet new needs. It is essential, therefore, that infrastructure supporting the industry is flexible enough to accept and absorb such changes with a minimum of disruption to the industry.

The nature of support for the sector must also be expected to change as the sector develops and matures. Thus, in the initial stages, and particularly where the industry is based on small-scale operators/producers, there will be heavy dependency on government inputs, such as funds, seed, feed, research, training, and marketing assistance, etc. Ultimately, with self-sufficiency and stability, government's role will alter so that its principal involvement will be regulating the industry within the wider socio-economic and environmental framework of the community. Its changing organization and management structure will reflect the reduced or new industry requirements. With increasing sophistication and complexity of a thriving industry, the need for a coordinating mechanism becomes more acute.

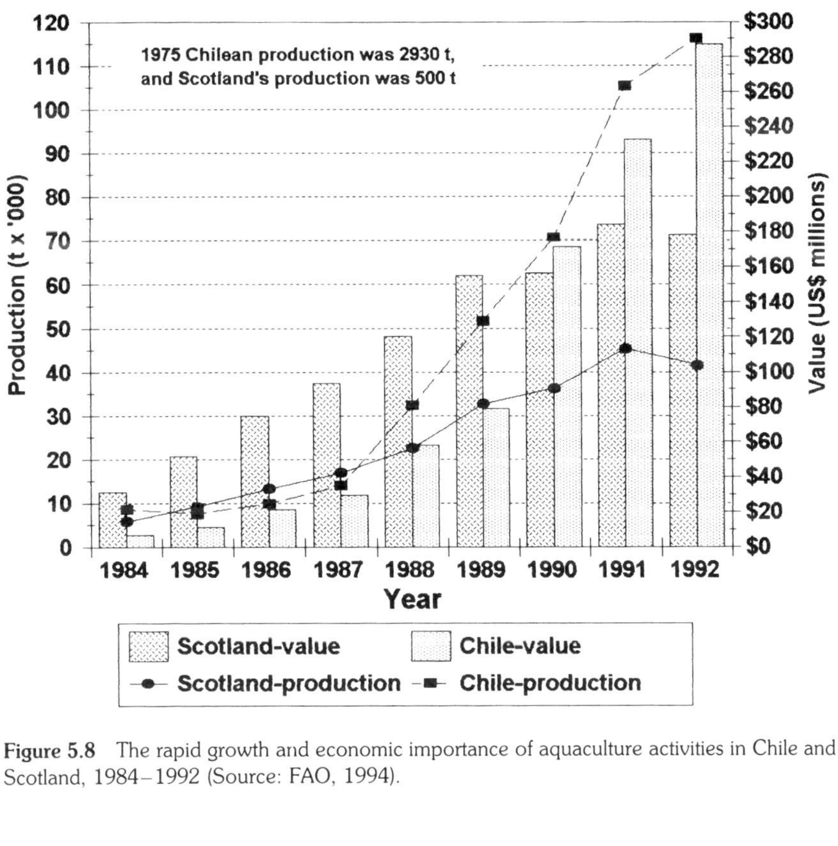

Figure 5.8 The rapid growth and economic importance of aquaculture activities in Chile and Scotland, 1984–1992 (Source: FAO, 1994).

As the sector develops there is a need for coordination

'In France, freshwater aquaculture is the responsibility of the Ministry of Agriculture; marine aquaculture is under the Ministry of the Sea. There is no clear demarcation in the responsibilities of the two Ministries for some aspects of land-based coastal mariculture or brackishwater aquaculture [and] other Ministries and regional and local bodies have certain responsibilities [and] must be involved. The Government of France decided to create a National Delegation for Aquaculture. The Delegation would have regional delegates based in the regional offices of either Agriculture or the Sea, depending on the structure of aquaculture in the region. These delegates are now the basic contact points for aquaculturists and potential aquaculturists. Their responsibilities are:

- to hand over any information needed to any administration concerned,
- to channel all applications for subsidies and soft loans to the right destinations, and
- to convene commissions of representatives of all administrations and institutions concerned to advise them of decisions.'

Source: ADCP/REP/89/34, FAO, Rome.

Thus, in many ways, the situation (quoted above) in France describes precisely what the end product of aquaculture development is liable to look like in terms of its organization and management. It also postulates one of the more rational solutions to an increasingly complex situation, though, of course, not the only one.

Chapter 6
The Use of Policy Instruments I – Legislation

Synopsis

The following three chapters (6–8) illustrate the range of policy instruments available and used by governments to achieve political and social objectives from the sector, and to direct and control its economic development. This chapter deals with legislation. It begins by reviewing the current place of aquaculture within the legal framework of countries, noting differences between the approaches and the extent and variety of existing laws and regulations. It comments also on the problem of the definition of aquaculture. The second section describes some of the legal requirements for setting up a farm, specifically the range of authorizations and in some cases qualifications required. The third section analyses the legislation governing access to and the use of land and water for aquaculture. The final section describes the environmental aspects of aquaculture which have necessitated legislation, particularly land use planning and zoning, operational control, discharge limits, fiscal incentives, and prohibitions. It finishes briefly with aspects of fish diseases and species movements.

6.1 Aquaculture in the legal framework

6.1.1 The place of aquaculture

Aquaculture, as described in Chapters 1 and 2, is a diverse and multidisciplinary activity. This diversity is also reflected in the variety of approaches from country to country in creating a legal framework for management. A review of aquaculture legislation by Van Houtte, Bonucci and Edeson (ADCP/REP/89/42) prepared as background information for aquaculture sector planning by ADCP, and summarized here, illustrated that there were many ways in which a state and its legal system could manage the sector. They confirmed that aquaculture, at the present time, does not yet have a firm legal status of its own, as it cannot be classified with agriculture, animal husbandry, or even fishing.

As aquaculture is a relatively modern technology it is generally assumed that most countries have not yet elaborated a legal framework in which it can operate. This is not entirely true. There are few legal instruments which date back to the origins of modern aquaculture in European countries in the late nineteenth century, and nothing in the legal records of their colonies into which it was introduced shortly after. It was left to the New World to produce the first real regulations. In 1883 a book was published in Mexico on the *Ideas on the importance of giving a vigorous promotion to Pisciculture and Aquaculture in the Country*. This led, a few years later, to the adoption by the Mexican Government of regulations to give legal foundations for expansion of fish culture, with provisions on application and petitions, authorization of establishments, transportation, and periodic reports.

It might also be assumed that customary land tenure practices in traditional societies would regulate early fish culture in some way. However, the lack of specific laws and regulations on aquaculture must therefore be attributed to other reasons. These are identified in the following sections, together with the needs of a legal regime to govern aquaculture activities.

Legislative policy instruments for aquaculture management are variable, and are not necessarily related to the existence and the development of an aquaculture industry. Aquaculture legislation can be divided, for all intents and purposes, into three categories depending on how countries have dealt with aquaculture so far. The three categories are as follows:

- Specitic sets of rules on aquaculture,
- Specific aquaculture legislation,
- Enabling law.

These are briefly described, with examples, in the following sections.

Specific sets of rules on aquaculture

Most of the western developed countries and some centrally planned states fall within this first category. For example, countries like Canada, the USA, New Zealand, Norway, France, Spain, and Ireland have a variety of legislation dealing directly with aquaculture and/or with certain forms it might take.

Consequently aquaculture in freshwater is considered by many states to be easily assimilated with a traditional activity, like animal husbandry, although marine aquaculture differs from capture fisheries so much that it must be treated separately. On the other hand, as inland capture fisheries and aquaculture are so interdependent that many countries have a single legislative text governing for both activities, and the basic principles of aquaculture are in a consolidated text in a law or decree on fishing. For example, in Hungary the text is a general framework of aquaculture legislation but also deals with open water fisheries, both commercial and sports fishing. However, in any land-locked country, such as Hungary and Nepal, it is always difficult to separate aquaculture and inland fisheries activities.

Some countries have a variety of legislation to cover many forms of aquaculture, and its activities

The legislation in France distinguishes between aquaculture carried out in freshwater and marine waters. In addition to this basic distinction, it has texts dealing explicitly with aquaculture in general or with certain practices; for example, a salmon hatchery is identified as an '*installation classe*' (classified installation) by law, and therefore subject to special procedures. Similarly, different regulations deal with shellfish rearing and culture, particularly from the environmental point of view. Freshwater aquaculture is embodied in the Rural Code, which is a general and basic text dealing with most of the activities taking place in the rural environment; marine aquaculture is regulated by a decree independent from the general Fisheries Law. A third environment is emerging for aquaculture offshore, and the need to regulate such activities in the open sea. The legal problems raised are different from traditional aquaculture in some important respects, and international laws may affect further development of this new practice. Presently France has no specific legislation, but legal implications of open sea enterprises are being studied.

The two sub-sectors are therefore managed in close connection, as they utilize the same natural and human resources, and have interdependent stocking programmes. Nevertheless, Hungary has also enacted several specific regulations on water use and supply for aquaculture purposes.

Many countries, although with different means and objectives, have recognized that aquaculture is a peculiar sector and important enough to justify its own set of rules. However they are in a minority, even amongst developed countries, and few developing countries, apart from some Asian states (such as the Philippines), have developed a comprehensive legal response even when aquaculture is dealt with in law.

Specific aquaculture legislation

Several states have a particular act and/or regulation to deal with aquaculture, but their numbers are few, considering that almost 170 nations now report some sector activity. Typically the Asian countries, as a group, have legal instruments but in most cases their existing legislation has broad application in that laws and regulations cover fisheries in general and not just aquaculture.

Some legislation, while not covering all aspects of aquaculture, is more specific with at least one legal text dealing with a certain type of aquaculture or a specific point related to it. For example, in 1975 the government of Ecuador, long before the country became the leading producer of farm-raised shrimp in the western hemisphere (with over 1300 farms, 100 hatcheries, 100 000 ha in production), enacted a regulation for the breeding and culture of aquatic species to control this potentially profitable sector. The scope of the regulation was to provide a legal framework for a precise type of aquaculture, namely, marine shrimp culture. In reality it established a legal regime for the use of coastal areas where shrimp farming would take place. Therefore the main intent of the text was to fix clear rules for the use of those coastal areas, which, being part of the public domain, were not open for any private ownership. While appearing comprehensive the regulation is limited to a certain type of aquaculture (marine shrimp farming) and focused on a specific question (access to the public land via the system of concessions).

Most countries which have adopted a specific act or regulation to deal with aquaculture have done so to respond to a precise need, like Ecuador. Others, and in particular developing states, have given some recognition and consideration to aquaculture by adopting some special legal instrument.

Most countries have adopted some legislation to deal with a specific need

In *Singapore* and in *Hong Kong*, taking into consideration their limited size, it was important to regulate access to water and land to avoid possible conflicts. In Singapore the Fish Culture (Control and Licensing) Rules of the Fisheries Act, provided for a legal regime for fish culture undertaken 'upon the seabed within the territorial waters of Singapore'. In Hong Kong the Marine Fish Culture Ordinance of 1980 designated 26 marine culture zones within which all marine culture activities could be undertaken; since 1983 some 50 offshore farms have been relocated into these areas. In *Ecuador* new legislation was a response to the uncertainty regarding the legal status of coastal areas (as there was a contradiction between the Maritime Police Code and the newly adopted Fisheries Act of 1974) and to stop illegal settlements on those zones.

An enabling law

Many countries have an 'enabling law'. This basic law is usually the Fisheries Act, which (a) sets up some principles on aquaculture, or (b) invests the legitimate authority with the power to regulate aquaculture. In both cases the basic legal instrument is meant to give a first input to a subsequent law but which, in general, is never elaborated further. The consequences of this lack of follow-up have been the impracticability, and therefore uselessness, of the legal provisions, and, in the worst cases with respect to aquaculture, supervention of ambiguities and contradictions between newly adopted basic law and the existing legal framework.

Basic laws including some provisions on aquaculture. A number of countries have provided for aquaculture in the form of a section or sub-section in the basic Fisheries Act.

Many basic national fisheries acts include enabling provisions for aquaculture

Provisions for aquaculture in Portugal are found in a General Decree regarding all waters under national jurisdiction. In Mauritania they are found in the Code for Marine Fisheries; in Congo in the Regulations for Marine Fisheries, and in Peru in the General Law on Fisheries.

Not all countries have achieved provisions the same way. For example, in 1974 the government of Colombia adopted an innovative National Code of Renewable Natural Resources and Protection of the Environment. The Code was divided into two books; one concerned the environment, the other devoted to the property, use, and environmental importance of renewable natural resources, which therefore included aquatic resources. Part of the Code defined the different activities of the fisheries and aquaculture.

Most of the Fisheries Acts which include special legal provisions on aquaculture are fairly recent. Most states now feel obliged to take into account aquaculture from a legal point of view without necessarily seeing any difference between culture and capture fisheries. However, the trend is not uniform, and recent marine fisheries legislation in Senegal and Mauritania either make no mention at all of marine culture, or contain no provisions on marine culture or other forms of aquaculture. Thus, notwithstanding the formal common pattern, the situation which has emerged is now quite different from country to country.

Countries with an enabling clause on aquaculture. Most countries, while not having any special law, section, or provision on aquaculture, have created the power to regulate its activities. The common approach is inclusion in the national Fisheries Act of an enabling article giving the minister or director of fisheries authority to regulate specific activities, including aquaculture.

This approach is widely used by developing states, especially in Africa. For example, in Kenya the enabling provision on aquaculture is included in the Fish Industry Act of 1968, according to which 'the Minister may make regulations . . . for regulating the manner in which any fish shall be caught, propagated, or cultivated'. This is similar to the wording in the 1974 Fisheries Act of Zambia, where the regulatory power granted to the Minister by the Act includes the possibility of issuing regulations for 'regulating and controlling fish culture and fish farming'. Malawi has adopted a similar provision in its Fisheries Act, 1973, where the Minister may make provisions for 'the regulation and control of fish culture, fish farming and fish breeding'.

Nonetheless, there are subtle differences between the approach of Kenya, and that of Zambia and Malawi. In the wording of the former, aquaculture is still considered to be an integral part of capture fisheries, and therefore the culture of fish is put on the same level as the catching and the stocking of fishes; in Zambia and Malawi the wording identifies fish culture as a field for which specific regulations can be enacted. More importantly, however, all three countries have not used the power to issue regulations on fish culture. The power is only potential, and so far it has been ineffectively exercised with respect to aquaculture.

These examples are typical of the attitude of most countries, and developing countries in particular, towards aquaculture (summarized in Figure 6.1). Many countries still do not have enabling legislation to deal with fish culture or any other form of aquaculture, despite the fact that it is becoming an economically important sector, nor do they make any direct reference to aquaculture in their national fisheries legislation. Nonetheless, this does not mean that it is not regulated in one way or another. As noted in the following sections, land laws, water laws, environmental regulations, etc., have regulatory and controlling effects. All these laws and regulations (summarized in Figure 6.2) are often decisive on the conditions for its development.

6.1.2 The definition of aquaculture

Aquaculture, as defined by the Concise Oxford Dictionary, is the 'cultivation of plants or breeding of animals in water'. This definition is broad enough to embrace descriptions such as freshwater culture, marine aquaculture, pisciculture, fish farming, sea farming, and perhaps even sea ranching. However, definitions of aquaculture usually raise more complex problems

- Few countries recognize different types of aquaculture in law.
- Many countries have legal provisions dealing with a specific form or a particular aspect of aquaculture.
- Some countries have the potential to regulate any activity.
- Aquaculture legislation is often linked to fisheries legislation, whether marine or inland.
- Current legal instruments are more responsive to *de facto* situations, than reflective on legal problems raised by aquaculture.

Figure 6.1 A summary of existing approaches to aquaculture legislation.

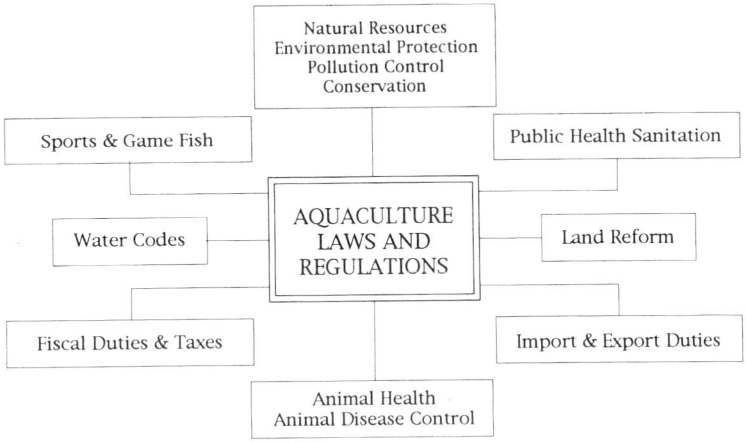

Figure 6.2 Groups of potential regulations and laws governing aquaculture.

than they resolve. Apart from the legal definitions of aquaculture and related matters, the definition of aquaculture for other purposes is not always straightforward. For example, the definition proposed by FAO, with its emphasis on ownership if only for statistical purposes, excludes several activities which would be included in the dictionary definition or, as will be seen, in most legislative definitions. This is because it is the nature of the activity itself, and not the elusive question of ownership, which is the central issue.

The definition of aquaculture by the Aquaculture Steering Committee of the Fisheries Department of FAO

'Aquaculture is the farming of aquatic organisms, including fish, molluscs, crustaceans and aquatic plants. Farming implies some form of intervention in the rearing process to enhance production, such as regular stocking, feeding, protection from predators, etc. Farming also implies individual or corporate ownership of the stock being cultivated. For statistical purposes, aquatic organisms which are harvested by an individual or corporate body which has owned them throughout their rearing period contribute to aquaculture, while aquatic organisms which are exploitable by the public as a common property resource, with or without appropriate licences, are the harvest of fisheries.'

In many countries, more commonly those with legislation drafted in accordance with common law principles, the principal legislation defines aquaculture. Where it is legally defined that definition will govern; where it is not defined, the term, or similar ones such as mariculture or pisciculture, are given their normal meaning unless the particular context requires that meaning to be adapted in some way. In the 1985 Fisheries Act of Malaysia, there are a number of terms defined which are relevant to aquaculture. Aquaculture itself is defined as 'the propagation of fish seed or the raising of fish through husbandry during the whole or part of its life cycle', while further definitions are provided of 'culture system', 'fish seed', 'fishing stakes', 'riverine fishing' and 'riverine waters'. The latter two terms make it clear that, included within them, in addition to rivers, lakes, and streams, are 'ponds and such other waters other than maritime waters, whether natural or man-made, privately owned or otherwise'. Here the comprehensive definitions of both aquaculture and its related issues provide a clear guide to the legislative intent.

In countries with legal systems based on the civil law model, definitions are less elaborate. Recognizing the difficulty of providing a perfect legislative definition, the Norwegian Act relating to the breeding of fish and shellfish states that 'for the purpose of the Act breeding means all activity where fish and shellfish are fed and handled with a view to consumption, feed production, reproduction, release, research or education'; but there is then added: 'in cases of doubt the Ministry shall decide what is to be considered as breeding'. Such a clause enables a resolution, at the administrative level, of any ambiguities and uncertainties which may arise with the definition.

Whereas the definition of marine fisheries in most fisheries laws occupies a key position in the legislative regime, the definition of aquaculture by itself (apart from any ambiguities as noted) does not advance matters very far. This is because the areas in which the activities take place are subject to much more complex regimes than marine fisheries. For example, as much aquaculture takes place in tidal areas (such as in estuaries and the foreshore of bays and inlets), or inland areas, land laws play an important role. Thus, in the absence of any specific definition in law relating to aquaculture, legal provisions which govern the delimitation of the foreshore (such as the determination of high and low water marks, and the definition of riparian rights), shape whatever legal definition of aquaculture may be provided.

Legal provisions for areas where farming activities take place often shape the legal definition of aquaculture

The 1959 Fisheries Act of Ireland enables the Minister to grant 'to the owner of any several fishery or the owner of the bed and soil of any estuary, a licence to form or plant an oyster bed within the limits of such several fishery or over such bed or soil above or below the lowest water mark of spring tides'. Here, the low water mark is defined precisely while the term 'estuary' raises some issues of geography (as not all rivers have estuaries) and legal interpretation, for in the context the term may cover all tidal areas of a river.

Another example of how an existing law may affect aquaculture indirectly is found in Guyana. The Sea Defences Act states that all sea defences (defined *inter alia* as all land fifty feet landwards from the centre of a sea or river wall and all land on the other side of the sea or river dam, to the toe of such sea or river dam or river wall), and future defences, are the property of the state. The Civil Law Act gives adjacent land owners on rivers title up to the median high tide.

Demarcation laws are almost as important to aquaculture as aquaculture laws themselves and, unless the latter contain clear provisions authorizing aquaculture activity, they probably have priority over aquaculture laws in particular legal systems. In addition there may be collateral questions in other legislation. In the United Kingdom, for example, where there are several different definitions of fish farms (depending on location, such as offshore or onshore, and the facilities proposed), planning permission under the Town and Country Planning Act is required for the development of land when operational or material changes are to be made. However, if the land is already agricultural in use, planning permission is not required under the Act if the purpose of development is 'the breeding and keeping of livestock, including any creature kept for the production of food, wool, skins, or fur' (Howarth, 1990). Hence as the end-product of most aquaculture enterprises is food for human consumption, they are frequently considered to be simply extensions of agriculture and subject to the same regulations.

6.2 Basic legal requirements for setting up a farm

States, as noted above, have given a variety of legislative responses to their emerging aquaculture industries, and it is necessary to examine more than the provisions which deal directly with aquaculture itself. Later sections (6.3 and 6.4) review questions of access and use of land and water, and the importance of environmental laws and regulations on aquaculture. Before this, however, it is necessary to describe some of the legal conditions for setting up of an aquaculture farm. This is different from any authorization, permit, concession, or licence required to have access to water or land.

In some countries these two authorizations are merged into one; in others, special provisions may provide for a formal linkage between them. For example, for activities implying the use of freshwater in particular, France provides a common procedure of public investigation regarding the rights to both the water to be used, and the fish; in Spain the Law on Marine Cultures provides concession for the use or discharge of seawater to be granted at the same time as the concession or authorization for the aquaculture enterprise.

Several states have established distinct procedures for obtaining an authorization to set up an 'aquaculture farm' – an expression intended here to cover all types of aquaculture establishments in all types of water, although the actual situation differs from country to country. There are both many commonalities and striking differences between these procedures. In particular:

- the scope of the law,
- the form and conditions of any administrative requirement,
- the general characteristics of it,
- the special limits and obligations attached to it, and
- the special rights it entails.

6.2.1 The scope of the legal instrument

With respect to the scope of the legislative act, and in particular the precise kind of activity it is intended to regulate, few countries provide for a single and general administrative regime. Only in Ireland does the scope of the Fisheries Act state that it applies to all types of waters (sea water, brackish water, and fresh water). In Mexico the Federal Fisheries Act applies to 'inland and internal waters, the territorial sea, the exclusive economic zone, and any man-made bodies of water', but it does not set up any administrative requirement, which is provided for in the General Regulation implementing the Fisheries Act. In El Salvador the General Law on Fishing imposes a common requirement (permit) applicable in all waters, whether public or private.

The fact that few countries have a single regime is due to two reasons. First, most national legal systems acknowledge a different legal status to water with respect to its public or private character; however, not all countries have the same definition of public and private waters as this depends essentially on their legal tradition (common law vs. civil law) and on the structure of the state (central vs. federal). Second, some states feel the need only to regulate certain types of aquaculture activities. For example, a number of states, such as the Philippines, Republic of Korea, and France, use the division between public and private waters to submit only the users of the former to administrative requirements, or to provide for different rules according to the legal nature of the waters utilized. Here, the administrative requirement covers both setting up of the establishment, and granting of a portion of water to be used for aquaculture. In other cases, states such as Hong Kong, Singapore, Spain, Mauritania, and New Zealand do not operate a division according to the legal status of the water but take into consideration its nature. Here, the local regime applies only to marine waters, whereas in other cases (France) only aquaculture installations set up on fresh water are concerned.

This division, however, is too rigid for all purposes and certain states, such as Ecuador, have enacted legislation which does not fit neatly into either category, and in New Zealand legislation dealing with freshwater

---

Legislation regarding farming in freshwater may be specific but still broad in scope

Extracts from fish farming legislation in New Zealand include:

- Authorizing persons, to be registered for the purpose, to construct and maintain ponds as fish farms for breeding and rearing fish (except trout) for sale subject to such conditions as are specified in the regulations;
- Regulating the operation of fish farms subject to such conditions as may be specified in the regulations, and prohibiting the operation of any such fish farm and the sale, disposal, transport, or export of any fish from such a fish farm without a licence;
- Providing for the registration of persons who may hold licences to operate fish farms and for the qualifications those persons shall be required to hold and the conditions with which they shall be required to comply before being granted registration and in order to remain registered;
- Providing for the appointment of officers, including any class of officers with specialist qualifications, to inspect such fish farms and ensure that any regulations made under this section are enforced and for the inspection of any such fish farm for facilities for processing or storing any fish reared on the farm;
- Providing for the issue, renewal, and transfer of licences in respect of any such operation and the form and contents of any such licence, and prescribing the fee, not exceeding NZ$300 per annum, payable in advance in respect of any such licence;
- Specifying the species, sub-species, varieties, or hybrids of fish that may be bred or reared in any such farm, and regulating the types and quantities of food that may be fed to any such fish, and the methods of feeding;
- Regulating or prohibiting the canning of any such fish, their sale or other disposal, or possession;
- Providing for the keeping by the licensee of any fish farm of records of fish acquired, kept, and disposed of, and for the keeping of records relating to those fish by any other licensee or other person;
- Prescribing the measures to be taken to avoid the outbreak or on an outbreak of any disease among the fish, and authorizing or requiring the taking of any specimen, the testing of any thing, or the sampling of any substance present on any fish farm and authorizing or requiring the removal of any specimen or sample, or the destruction of diseased fish, whether with or without payment of compensation.

Source: Ministry of Agriculture and Fisheries (New Zealand), 1983.

---

aquaculture farms deal explicitly and exclusively with 'fish farms'. Here, fish farm means 'any establishment (including an ocean ranching establishment), licensed under these regulations to capture, raise, feed, release, process, or deal in fish for sale; but does not include:

- any hatchery established and operated by an acclimatization society or by the Ministry of Agriculture and Fisheries or by the Department of Internal Affairs,
- any food premises, or
- any establishment licensed under these regulations solely to process or deal in fish for sale'.

The legislation in New Zealand is interesting for what it includes and what it excludes. In particular it includes ocean-ranching establishments, and defines ocean ranching as, 'releasing juvenile salmon into waters to grow at sea and, on the return of the adult salmon to a specified place, harvesting them or producing ova from them'. The extent of the regulation is therefore much wider than suggested by its title.

6.2.2 Form of the administrative requirement

Although the scope and extent of the parent Act vary from one country to another, some common patterns are apparent. The most striking similarity, from the examples above, is the need to obtain authorization from the administration before setting up aquaculture facilities. This authorization may be granted by different authorities, and may have different names, but still all countries exemplified above require approval from some relevant public agency. In contrast, in the agricultural sector many activities do not have to be authorized, and no country uses the system of notification.

The authorizing authorities

Authorizing authorities are often either the central administration, usually represented by the minister of fisheries or someone acting on his behalf, or a decentralized authority at either a regional or municipal level. Where different interests are involved, no authorization is usually granted without prior approval of several other authorities.

Special situations often occur in countries with a federal or decentralized form of government. A number of states, such as Canada and the USA, have a federal structure which distributes authority between the central government and state or provincial authorities. The consequence is that this further complicates the legal regime for aquaculture. For example, in the USA there are about 120 federal laws, scores of state and municipal laws, and hundreds of regulations which have an impact on aquaculture. Federal aquaculture laws generally fall into two categories: those which apply to waters located beyond state jurisdiction, and those which authorize

---

Different authorities may authorize aquaculture activities

- In *Singapore*, according to the Singapore Fish Culture Rules, 'every person working on fish culture farms must be registered by the Director'.
- In *New Zealand* under the Freshwater Fish Farming Regulations, 'no person shall establish or operate any fish farm except pursuant to a fish farm licence granted by the Minister under these regulations and for the time being in force and in accordance with the conditions of the licence and with the provisions of these regulations'.
- In *France* all requests concerning a freshwater aquaculture establishment must be addressed to the Prefect of the local Department (District).
- In the *Republic of Korea*, according to the Fisheries Law, those wishing to operate culture fishing 'shall obtain licences from the Mayor of the Special City of Seoul, the Mayor of Pusan City, or provincial governors'.
- In *Ecuador* aquaculture activities may be undertaken only under legal authorization issued by the Directorate General for Fisheries. When these operations are carried out in coastal areas, a special licence from the Ministries of National Defence and of Natural Energy Resources is requested, in addition to the authorization.

---

conduct of research and development programmes and construction and operation of federal hatcheries. None of these federal laws or regulations applies directly to private farming of fish, shellfish, or marine plants. Other federal laws, related to land and water use, environmental protection, and health or safety, also affect private aquaculture ventures. Many of the laws and regulations which specifically authorize permit, control, or prevent aquaculture are at the state level. State laws concerning aquaculture are usually closely related to state authorities charged with management of fisheries resources. Hatchery programmes are therefore usually administered by state fish and game, or conservation agencies. The procedures to obtain permits and licences have been a severe deterrent to aquaculture in the USA.

Similarly, because aquaculture operations deal with food production, water supply, the use of navigable waters, and effluent discharge, there are regulations for other agencies concerned with wholesomeness of food, public health, water purity, land use, and pollution control. Such regulations may apply for all three levels of government – namely federal, state, and local levels.

Together with these federal states there are countries with a high level of decentralization. In these countries special laws have devolved a certain

Plate 6: Aquaculture facilities must comply with regulations regarding local navigable waters. Plate 6a (top) illustrates approved development for pearl oyster culture in Mie Prefecture (Japan); and Plate 6b (bottom) shows licensed salmon farms close to the car-ferry service from Seattle along the coast of Bainbridge Island, Washington (USA).
Photos: Courtesy of Conrad V. W. Mahnken.

number of competences onto regional administrations. In some cases, such as in Italy and Spain, this can give them authority to regulate not only inland fishing activities but also marine fisheries within the territorial sea.

The form of the authorization

Most countries, as noted above, have an administrative procedure based on prior approval by the administration for setting up aquaculture facilities. One alternative is in the Act for Breeding of Fish and Shellfish in Norway. While submitting the breeding of fish to a general obligation to obtain a licence, one section stipulates that, 'a licence . . . for the breeding of shellfish and species of fish other than salmon and trout *shall* be granted provided that this is not prevented by the provisions in (other) sections'. This section, which was intended to favour the development of species other than salmon or trout, had the same effect. However, success was so great that it is now feared that the farming of salmon and trout might suffer. Therefore, in a Report on Aquaculture (1987), it was stated that 'The Ministry of Fisheries will present a proposal whereby an amendment is made to the Act implying that it no longer contains a provision stating that licences for the farming of other species must be granted'.

The names under which any authorization is given are different, and the following are widely used:

- *authorization and concession* (in France, Congo, Spain, and Ecuador),
- *licence and permit* (in Republic of Korea, Philippines, Norway, Ireland, Hong Kong, Singapore, and New Zealand), and
- *lease.*

Although these terms sometimes are used in distinction to each other, in reality they amount to the same for the aquaculture sector.

Usually the authorization which is being granted is a licence to culture. The 1967 Fisheries Law of Korea qualifies a licence with the words 'for cultivating marine animals or marine plants in certain demarcated water or work facilities', but most countries offer broader scope in the licensing system.

In most countries a licence for aquaculture provides broad scope for interpretation

Regulations for farming in Norway state, 'without a licence from the Ministry, no person may build, fit out, extend, acquire, operate, own or possess a majority interest in a facility for the breeding of fish or shellfish. Nor may anyone, without a licence, carry out breeding by a method other than a breeding facility'.

Most countries have a universal licensing system to include any species intended for culture. However, on some occasions special licences or authorizations are required. In the Philippines, for example, together with the ordinary licence a special licence is necessary to construct or operate a fish pen. Similarly, specific regulations are in force concerning the culture of mussels, the culture of seaweeds, and fish in pens and cages. Ireland there is a peculiar licensing system due to the existence of two basic Fisheries Acts. Accordingly, the 1980 Fisheries Act provides that it shall not be lawful for any person at any place to engage in aquaculture without one of the following:

- a fish culture licence,
- an oyster bed licence,
- a licence granted by the Minister under this section,
- an oyster fishery order.

France requires a special authorization for salmon culture facilities, whether the activity takes place in public or private waters, in sea or freshwater. The requirement is embodied in a law on 'classified installations for the protection of nature', in which are considered buildings and/or constructions, which for their possible impact on public health, on agriculture, on conservation of nature and the environment, had been put on a special list. Salmon culture activities were put on the list in 1983. In this respect the authorization to set up an aquaculture farm can be linked to obtaining other types of licences (however named), for example, access to water and/or land, etc. In most cases this is not explicitly provided for in the legal instrument relevant to aquaculture; nonetheless, New Zealand demands that, prior to the grant of an aquaculture licence, the applicant 'shall obtain under the provisions of the Water and Soil Conservation Act 1967 any right required by the Act in respect of any water required for, or affected by, the proposed fish fauna'.

New Zealand has another particular feature, which is to contemplate two varieties of freshwater fish farm licences for both production and selling the products. According to a section of the Freshwater Fish Farming Regulations:

- No person shall establish or operate any fish farm except pursuant to a fish farm licence granted by the Minister under these regulations, and
- No person shall process or deal in any produce of a fish farm except pursuant to a fish farm licence, or a licence to process or deal in the produce of a fish farm, granted by the Minister under these regulations.

6.2.3 General conditions attached to the authorization

The following section describes the principal terms and conditions attached to the authorization.

Obligation to provide documents and information

Several countries require applicants to provide the granting authority with a set of documents (or a '*dossier*' in France) containing information on the intended activity. The amount of information required, and the degree of detail, vary from country to country. Many countries have limited requirements even for a special type of aquaculture, such as the culture of mussels.

Most countries have limited requirements for aquaculture licences or permits

A section of the Rules and Regulations of the Philippines governing the culture of mussels states, 'Application and requirements: A duly accomplished application for the culture of mussels shall be filed with the Provincial Fishery Office where the area is located and accompanied by the following:

- a sketch plan of the area applied for in the designated mussel belt,
- an application fee of P20.00, either in cash or postal money order payable to the Director which is non-transferable and non-refundable, and
- a xerox copy of the duly approved Articles of Incorporation and/or Certificate of Registration, when applicable.'

In contrast, developed countries, with either common law or civil law traditions, have more detailed requirements. For example, the Freshwater Fish Farming Regulations of New Zealand require detailed plans showing the location of the proposed fish farm and all premises associated with the farm for processing or dealing in its produce, layout of all ponds and buildings, all intake and outlet structures, where applicable the source of water to be used, flow pattern of the water through the fish farm, and location of outflows, and, in relation to ocean-ranching, the place from which salmon are to be released and the site of recapture. Also the applicant shall specify his right to occupy the area of the proposed fish farm, and shall supply sufficient evidence that all rights, licences, and other authorities required by any enactment authorizing him to draw water from and return water to any lake or river or stream, to take or discharge water into the sea, and to occupy any foreshore have been duly granted and are in force, etc.

The regime of the French legislation is even stricter. In the Fish Culture Act the dossier must include certain identified documents and plans to a specified scale, including (among others) the nature of the fish farm, a report describing the nature and proposed methods of fish rearing, the fish species, the production target or purpose of the experiment and the method of harvesting the fish, the emptying schedule, and the period for which the authorization is requested, the period foreseen for completing any arrangements, and a note indicating the financial capacity of the applicant in relation to the planned undertaking.

These three examples, together with the relevant provisions of other states, permit some conclusions on the type of information usually requested by states. These are summarized below.

Applicants for licences are usually required to provide three types of documents and information

- *Economic and administrative information.* This may include names and legal status of the applicant, his financial capacity, duration of authorization requested, any relevant title, etc.
- *Geographic information.* Including, *inter alia*, a map of the requested area with, in some cases, a detailed design of the premises and facilities.
- *Technical information.* The species to be cultivated, methods to be adopted in operating the aquaculture farm, the production target, etc.

Citizenship, professional qualifications, and other criteria

In several cases applicants for licences to practise aquaculture must fulfil conditions of citizenship, and, in some (mostly developed) countries certain professional qualifications are required.

If the legislative text on aquaculture takes a stand on the question of citizenship it is usually in favour of its nationals. Such restrictions are not necessarily limited to developing countries. In France, for example, the possibility of operating a mariculture farm is available only to EU citizens, and Spain allowed only citizens to undertake aquaculture until such provisions were modified on accession to the EU (formerly the European Community, EC) in 1986. There are probably several reasons for such restrictions on access of foreigners to practise aquaculture, the most important of which is that aquaculture implies a prolonged occupation of water and land.

For restrictions regarding professional qualifications and other priority criteria, France is the leader. In France the requirement for professional qualifications is intended to ensure the most appropriate development, particularly in marine culture in French waters. The government recognizes that marine culture entails occupation of a part of the sea and, notwithstanding its economic importance, it must co-exist with other vital activities. Thus traditional capture fisheries and free navigation are considered, together with potential risks to the environment caused by aquaculture.

A few countries have set priorities for granting authorizations for aquaculture enterprises to achieve other objectives, and not only through the professional qualifications of the applicants. In Norway selection criteria fulfil general objectives of economic and social efficiency. The wording is loose and gives large discretionary power to the granting authority. Other criteria are more precise, recognizing priority rights to local interests. Both these provisions contain clear aspects of policy.

Few countries, and virtually no developing countries, have yet established a fully comprehensive and financed plan for the development of their aquaculture sectors. Nevertheless many of them have adopted practical provisions which, like Norway, are intended to influence development of the sector in certain directions, and they recognize the diversity of options available to achieve the same purpose.

Several governments plan direction and control of the sector through regulatory measures

Provisions typical of aquaculture include:

- granting special rights to public administrations or public entities to operate aquaculture farms,
- granting special rights to cooperatives or associations to operate aquaculture farms, and
- granting special rights for the development of a certain type of aquaculture.

Duration, payment, renewal of the authorization

Granting authorization is a necessary step to undertake aquaculture. However, as with the development of all sectors, the terms and conditions attached play an important role in stimulating interest of the people. In particular, if aquaculture is not a traditional activity it is important to provide for an attractive regime, including the fiscal and financial incentives which can be given (see Chapter 7). For promotion of the sector, the duration of the authorization, its cost, and the possibility of its renewal, are all elements which both the government and the private enterprise bear in mind.

Duration of the authorization. The name under which an authorization is granted, as noted above, is not very important. The main purpose is to guarantee sufficient stability for the investment by the beneficiary of the authorization. States generally grant an authorization for several years. In the Philippines, for example, a licence for the culture of mussels is valid for three years, and a licence for the construction and operation of fish pens is for five years. For fish ponds the original legal regime first instituted in 1975 provided for a one-year permit or a 10-year lease agreement, but in 1979 the Government allowed for the conversion of these authorizations into a 'twenty-five year fish pond lease agreement . . . otherwise said permits and lease agreements shall be deemed automatically terminated/cancelled'. In

Spain a concession for marine culture is granted for 10 years, and in France 35 years.

Another approach used by some states is to give discretionary power to the granting authority. This is the case in Hong Kong, where no period is set; and also for freshwater fish farming in New Zealand but, although no actual duration is specified in the licence, it cannot exceed 14 years.

Payment. The amount to be paid for the authorization is also important. It may differ within each country for each specific production activity.

Different licences often have different fee structures

For example:

Ecuador	annual fee	ES300 000 per hectare authorized for culture.
Singapore	annual fee	S$500 for each 5000m^2 and S$5 for every person working on a fish culture farm.
Philippines		
• seaweed	not specified	(a) P20.00 for the application fee to culture, (b) P50.00 per 10 000 m^2 or fraction thereof, for the licence fee to cultivate;
• mussels	annual fee	P50.00 per half hectare of fraction thereof payable in advance;
• fish pens	annual fee	P200.00 per hectare or fraction thereof;
• fish cages	annual fee	P20.00 per 100 m^2 or fraction thereof.

A number of states, in particular those under civil law (such as France) using the system of concession, indicate the amount of payment in the contract of concession. For those under common law (such as New Zealand) the amount of payment appears in the schedule. In Norway the regulations are ambiguous regarding payment at all.

It is difficult to form any conclusion on the level of fees because the countries, the type of cultures, and the time the fees were fixed, are different. However, from a brief comparison some common patterns can be observed. First, the fees have to be paid annually even if the authorization is granted for several years (this is the case for Ecuador and the Philippines); second, the criteria of the amount of land occupied and water used by the farm are key factors, and third (as in the Philippines), the type of culture may be taken into account.

Renewal of the authorization. From a comparative analysis of the current practices in different countries, a licence is usually renewable under the same terms and conditions under which the first one was granted. This includes in particular its duration. Such provisions are found in the Philippines, Republic of Korea, El Salvador, New Zealand, and France. A peculiar provision is found in Spain regarding marine culture; according to the law, whilst the first concession is granted for 10 years, its renewal may be authorized to a maximum of 50 years.

Special obligations and rights

Many countries attach certain obligations and rights to the granting of a licence to practise aquaculture, some of which are quite diverse.

Special obligations. The majority of countries provide for specific obligations on behalf of the enterprise. In some cases the obligations are stated clearly before granting the authorization.

Authorizations are often granted with obligations

In France, 'No fish farming authorization or concession may be granted if it appears that any adverse situation is likely to supervene for the fish living in any waters connected with the fish farming establishment in question, *inter alia*, if the installation were to lead to an interruption in the unrestricted circulation of the fish in the watercourse, to an insufficient flow-rate, or to an impairment in the quality of the water such as to place the lives of such fish in danger. Furthermore, no authorization or concession may be granted unless the proposed fish harvesting procedure excludes line fishing.'

In most of the countries, however, the main requirements are focused on the subsequent use of the authorized area for culture. Different kinds of obligations may therefore be provided by the law. For example, there is frequently an obligation to give regular information and to grant permanent access to the site (as in Norway); in the Philippines people practising the culture of mussels must submit a half-yearly report of production for statistical purposes. Another common obligation is that the enterprise must exist and operate for a stated duration; in El Salvador law states that if the area is not exploited for six months the permit will be cancelled. Frequently there is an obligation to improve the area granted; in the Philippines the 'licensee' is required to introduce improvements in the area covered by the licence within 60 days from the date of issue. Finally, there is often an obligation to mark or delimit the concession; in Singapore 'the licensee shall display on the fish culture farm such identification mark as may be prescribed by the Director to denote that it is a licensed fish culture farm.

The licensee shall ensure that his fish culture farm is lighted at night to the satisfaction of the Director.'

Naturally these specific obligations are all in addition to general requirements of respecting the existing legislation, and the terms and conditions specified in the authorization.

Specific rights. If an enterprise must fulfil certain obligations it may in some cases also be given specific rights. For example, there may associated rights for:

- transferability or cession of the authorization,
- fishing, or
- other legal protection.

The right of transferability or cession of the authorization is a delicate question for the authorities. On the one hand a certain guarantee must be offered to the beneficiary; but on the other, particularly if the enterprise is in the public domain, the government must reserve its rights. Consequently there are a variety of solutions in legislation. Apart from legal instruments which consciously or unconsciously take no stand on the question, in others the possibility of transfer is clearly prohibited. For example, transfer is often forbidden (as in Hong Kong and Ecuador), as is the sale of the authorization (Ecuador). However, many countries realise that such rigid prohibitions constitute an obstacle to the development of the sector and allow the possibility of transfer and/or cession of the title.

Most countries include rights to transfer authorizations to promote development of the sector

In the Freshwater Fish Farming Regulations in New Zealand regarding authorization of a licence: 'Any person, being: the licensee; the personal representative of a deceased licensee; the manager under the Mental Health Act 1969 of the estate of a licensee; in the case of a licensee in respect of whom a protection order is in force under the Aged and Infirm Persons Protection Act 1912, the manager of his estate; the assignee in bankruptcy of a licensee who has been adjudged a bankrupt; or the liquidator of a company which is a licensee – may transfer the license to any other suitable person. However, the transfer must be approved by the Minister, provided that, 'he is satisfied that the proposed transferee, having regard to his qualifications, experience and resources, is a suitable person to operate the fish farm or processing plant; and provided that the Minister shall not approve any such transfer unless he is satisfied that all licences and other authorities required by any other enactment authorizing the proposed transferee to draw water from and return water to any lake or river or stream, to discharge water into the sea, and to occupy any foreshore have been duly granted and are in force.'

In Spain the legislation on marine culture includes the possibility of transferring the concession or authorization. In this case the authorization of the granting authority is requested and, if the beneficiary is more than one person, it is given on behalf of the group. France has also set up a system for the transfer of a marine culture concession. The law permits the 'substitution' (transfer) of the concession to a member of the family or even to a third party under the following conditions: the concession must be held for 10 years if the transfer is to a third person or for five years if it is a family transfer; furthermore, the beneficiary must fulfil the same legal requirements as the original licensee. However, 'substitution' can also occur on payment of a sum which takes into account not only the facilities constructed but also production of the enterprise, thus making the transfer similar to the sale of a shop or a factory. In other words 'substitution' is a quasi-property right.

Fishing rights are not usually explicitly mentioned in the law. One exception is the 1967 Fisheries Law of the Republic of Korea, specifically a fishing licence carries with it fishing rights which are 'under the category of right in rem, and legal provisions pertaining to land shall apply, with all necessary changes, to the same'. In Norway the authorization recognizes the right to 're-catch' fish on the following grounds: 'It is prohibited for any person other than the owner of the facility who has a licence in accordance with Sub-section 3 to re-catch fish or shellfish which are no longer penned in and are to be found in a free state in the vicinity of the facility. The right to re-catch may be exercised for up to 14 days after escape; this applies even if a period has been fixed when the equivalent species are protected'.

Finally, as aquaculture enterprises frequently conflict with other activities and interests, some legislation recognizes this and provides particular protection. In Spain, for example, in zones declared to be of interest for marine culture all activities which can have an effect on the quality of water must be equipped with adequate protection systems. El Salvador goes further in providing that in the areas where aquaculture is practised, no navigation, fishing, or straying of unauthorized persons is permitted.

In summary, it is evident from this review of the basic legal requirements for setting up aquaculture enterprises that farming activities are not easy to regulate. This is not only because of their diversity, but also because of their interdependence on other basic aspects of any national legal system, particularly laws regarding land and water use. This is the subject of the next section.

6.3 Access to and use of land and water

Aquaculture takes up land (some of which may be under water), uses water, deals with animals, and may require a variety of chemical and veterinary medicines in its operations. This section reviews and analyses the legislative policy instruments governing access to and use of:

- land for aquaculture, and
- water for aquaculture.

6.3.1 Access to and the use of land

In many countries little attention is paid to the use and access of land devoted only to aquaculture purposes. But the laws regarding private and public property rights, agriculture development policies and land reform Acts (insofar as fish culture is regarded as a branch of agriculture), environment protection policies, and finally national security and defence considerations, are those most likely to regulate access and use of land for aquaculture enterprises.

Publicly owned land versus privately owned land

Aquaculture activities are invariably carried out in coastal areas or on riparian lands, and in some cases on forest lands. Access to and use of the sites, where they fall within the public domain, are most likely to be subject to special concessions or authorizations granted by competent authorities. In France and Ecuador, for example, special regulations provide that activities located on the public maritime domain may only be carried out within an authorized concession. In both these countries the procedure aims to obtain, during the 'inspection' phase, a favourable report of all persons and state organizations concerned, and to involve various ministries in allocating authorizations.

Licences, permits, or lease agreements for the use of the public domain for aquaculture purposes are also in force in several Latin American countries, such as Peru, Honduras, Colombia, and El Salvador, and are common to different countries in East Asia. These usually cover the use of the public domain for any aquaculture purposes, such as fish pond lease agreements or leasehold rights, or to give holders of fish-pen permits and licences to operate concessions in fry collecting grounds, among others. Much of this legislation takes into account an essential aspect of fish farming, namely the importance of long-term leases for aquaculture to allow for the recuperation of long-term investments. Moreover, exploitation rights of any other resources within the area covered by the lease which may be detrimental to the interests of the permit holder or lessee, are not granted to other persons.

In some Pacific Island nations, reef and lagoon areas are owned by villages or families. In these cases aquaculture operations must meet with the approval of the appropriate chief or family head. These traditional management systems are still operational in the Cook Islands, Palau, Federated States of Micronesia, and other island states.

Legislation concerning public lands directly influences the access and use of them for aquaculture purposes. In most countries it seeks to offer equal opportunities for nationals, whether administrative units (as in Peru), small-scale fisherfolk groups or family sized groups (Philippines), or large corporations (France).

Legislation concerning the use of public lands seeks to offer equal opportunity for nationals

Foreign investors and assistance (foreign capital and technology) are welcome in most countries, as long as national investors are not adversely affected, control of the sector is not lost, and nationally oriented sector developments are not hindered. In the Philippines and Ecuador, for example, access to sites is subject to a complex administrative procedure and foreigners are allowed limited access only. In France the 1983 legislation for marine and coastal aquaculture was a strong protector of small-scale national enterprises and made it extremely difficult for (multi-national) corporations to have access to aquaculture sites. In 1986, the legislation was revised in France to facilitate transfer of aquaculture enterprises from one owner to another, and gave access to corporations for investment.

Agriculture development policies and land reform regulations

Many policies and laws influence access and use of land for aquaculture purposes; this is particularly true in Colombia, Panama, and Venezuela, and to a lesser extent in Tunisia, Zambia, and the Philippines. In accordance with national development programmes, areas suitable for aquaculture were granted more easily under the form of leases to small-sized groups, individuals or corporations, particularly for the development of uncultivated land for agricultural and pastoral purposes. The beneficiaries have had to abide by the existing rules for the exploitation of certain areas, such as coastal mangroves, delicate ecosystems, and places with other types of renewable natural resources.

In a different way the Fisheries Law of Egypt states that the establishment of fish farms is prohibited except on fallow land that is unsuitable for agriculture; and existing agricultural lands and farms cannot be prejudiced. An authorization from the Ministry of Agriculture is required.

Protected areas

Legislation regarding so-called 'protected areas' for aquaculture operations, carried out by private or public bodies, frequently hinders or facilitates access to and use of lands suited to aquaculture. Various reasons justify such rules, such as environmental protection (France), aquaculture protection and promotion (Singapore), natural resources management, educational, research and scientific development purposes (Philippines), public health protection (Philippines), management of the conflicting uses of certain natural resources (USA), and protection of water quality.

Legislation for protected areas can both facilitate and restrict aquaculture development

In Ecuador the establishment of shrimp hatcheries along the coast is only allowed in technically-specified areas. A Commission, formed by the Oceanographic Institute of the Army, the Directorate for Merchant Marine and Coasts, the Directorate General for Fisheries, and the National Fisheries Institute, decides on the location of the specific areas. The competent authorities have refused to recognize, and consequently legalize, some existing shrimp hatcheries because they were not located in these areas. The authorities have the power to direct that the areas be returned to their natural state.

In France the situation is even more complex and traditional town planning documents and maritime area development documents are used as tools to preserve and protect potential sites. These documents provide for the optimization of the various activities located there, as well as resolving problems of compatibility between different uses. In Zambia the establishment of special areas is motivated by the wish to protect or control the natural resources, and quality and quantity of water resources.

Finally, in some countries intervention and interference of the government for purposes of national security and defence are likely to limit or hinder the access to and use of certain lands. In Ecuador the Ministry of National Defence and the Joint Command of Armed Forces are regularly involved in the decision-making process.

6.3.2 Access to and the use of water

Water, together with land, is one of the major and basic natural resources for which each government tries to plan orderly economic policies at national, regional, and local levels. In most countries, the legislation on water resources deals with its beneficial uses (domestic, municipal, irrigation, etc.) the harmful effects of water, the uncontrolled waste and misuse of water, and finally the quality and environmental aspects. Each of these subjects often seems to be administered by a different ministry, department, or autonomous institution without apparent coordination.

Most countries have a regulation dealing with the right to use water resources for aquaculture purposes, either specifically or as a part of the right to use water for fishing or agriculture purposes in general. In the same manner regulations concerning discharge of waste water and effluent, and more rarely the quality, can be found. In some countries, such as Panama, the law is concerned first with conservation of water resources, or their optimum use, and only then with other uses. In other countries, such as Colombia and Poland, the law deals with water in general, together with other natural resources and environmental elements.

Laws regarding water for aquaculture may be general in nature

Priorities are determined on a regional basis regarding ecological, economic, and social considerations, allowance being made for the need to establish reserves, the wish to conserve the environment, the economic and social costs and benefits of any project, and considerations of subsistence and the economic and social development of the population of the region concerned. Furthermore, like any other constituent part of the environment or renewable natural resource, water must be used in an efficient manner so that the maximum gain may accrue, compatible with the general interest of the community and with the principles contemplated by the National Code of Renewable Natural Resources and Environment Protection, in such a way that no interference shall occur with any of those resources or with the rights of third parties.
Source: Ministerio del Medio Ambiente (Colombia), 1974.

Whether adhering to water law principles through common law or civil law, or other systems, the amount of water available for aquaculture purposes is almost invariably stated in a permit, licence, or authorization from local, regional, or central governing bodies. Differences in the following fields are most likely to influence the regime of water use for aquaculture purposes, namely:

- the legal status of ownership of water,
- an explicit reference to the need of a licence for aquaculture, and
- the place of aquaculture in the order of priority established among beneficial uses, if there is priority.

Because of the different conditions existing from one country to another and the limited amount of information available, it is difficult to summarize the existing situation with respect to water resources administration and legislation for aquaculture purposes. However, several countries have a statute on water use for aquaculture purposes (for example, Mexico, Guatemala, Poland, and France). Others have a number of discrete enactments which either have an indirect bearing on water use and control on the subject (for example, UK and Zambia), or which cover only certain aspects of water management without reference to aquaculture as such (Malawi and Panama). Most of the latter countries acknowledge the right to use water for agriculture, animal health, etc., and the point is to know whether or not fish culture can be understood as such.

In addition, even in those countries with laws referring to aquaculture, it is rarely specified if authorization granting the right to use water for culture has a temporary character or not, or if the duration of the water licence agrees with the duration of the aquaculture licence, if any exists. At the same time, it is also not known whether the duration meets or is adapted to aquaculture needs. The only known countries where the aquaculture licence is issued in complete accordance with the water laws are El Salvador, Guatemala and France. In El Salvador the General Board of Fishing Resources must grant a permit to engage in aquaculture operations in accordance with the law on uses of water (General Law on Fishing Activities).

The priority given to the use of water for aquaculture purposes is not always clear in those countries where a priority order exists. It is most often placed after domestic use, and use of livestock watering, or for public interests. It is sometimes preceded or followed by the use for agriculture purposes and usually followed by the use for industries or recreational purposes, etc. (depending on whether aquaculture is included in one or another category). Obviously water management policies are to optimise the use of water (for domestic and drinking purposes) before meeting other needs. However, the criteria for classification are not always clear. Lack of awareness of the importance of regulating the use of water for aquaculture could be one reason. It is true, however, that aquaculture operations have been integrated in a legislative framework only recently in most countries, whereas water laws are long established.

Finally, only in France is there a different legal statute between the use of seawater and the use of freshwater for aquaculture purposes (bearing in mind that most countries adhere to the principle that all waters vest in the state). It appears that no significance is attached to the various types of existing aquaculture techniques, i.e. brackish, fresh and marine water fish culture.

6.3.3 Discharge of used waters and water quality

Aquaculture, like agriculture, has the potential to cause environmental impacts on rivers, lakes, and coastal areas, primarily as a result of the discharge of wastes. Conversely, water pollution has the potential to impact on aquaculture enterprises (see section 2.1). Laws dealing with water discharge, water quality, and environment protection are now to be found in almost all countries. Pollution prevention regulations, water use laws, natural resource conservation laws, health codes, and several other groups of laws regulate all these matters.

Water discharge

From the perspective of aquaculture enterprises, legislation on water discharge varies from country to country. Not all legislation, however, deals specifically with the discharge of used water by the farms, but the general requirement that discharges of 'pollutants' must be authorized. In other words, the important common feature is the requirement of prior authorization for discharge of 'waste' into water. The authorization is either incorporated in the grant (permit or concession) to use water, as in Zambia and Ecuador, or is given separately, as in the Republic of Korea, Malawi, Poland, and the UK.

Many countries deal specifically with discharge of water by farming enterprises. Aquaculture-related laws and discharge rules might be part of a single package, insofar as the aquaculture concession or permit creates reciprocal obligations between the user and the administration, and among them there exists the obligation for the farmer to provide suitable installations or equipment to discharge the used waters to conserve its quality. This is the case in New Zealand, France, and El Salvador. Thus, freshwater fish farmers are not able to start activities without prior authorizations to (a) create such a farm, (b) abstract and discharge water, and (c) and protect the environment.

Finally, in other cases, discharge of used waters by fish farms is often covered by a specific regulation. France again, together with Hungary and Ecuador, are countries in which a particular regulation exists regarding the discharge of water by particular types of farms. Shellfish farms, for example, and more particularly those equipped with oyster depuration facilities, are subject to a special regime: the waste waters must in all cases be discharged in a sewerage system and never into natural waters.

All these rules place a considerable amount of power into the hands of water authorities to determine whether a farming enterprise should be conducted or not, and if so at a particular site.

Water quality

The importance of water quality and its protection, for aquaculture purposes, has not received much attention. Environment protection laws, water laws, natural resources protection, and conservation laws deal mostly with water pollution control regulations, either in a general or specific way, or with any aspect considered separately. More general regulations might prohibit pollution in leased areas, or where there are several plants in the same area, or create civil liability for any damage or consider policies on marine pollution control.

Few countries make any real distinction between the control of inland water pollution and coastal water pollution. There is some question if this is justified for the actual quality standards. Within the EU, Council Directives have been made in order to protect, preserve or improve the quality of waters relevant to aquaculture. The waters concerned are salmonid and cyprinoid waters on the one hand and shellfish waters on the other (see Figure 6.3). Pursuant to both directives, quality objectives are met by fixing, first, parameters to measure the quality of the waters and, second, by establishing a set of reference values. The competent authorities in the member states must carry out sampling operations and those samples should be in conformity, for at least a certain period (one year), with standards set by member states.

The protection and improvement of the environment necessitates, according to the Council, concrete measures to protect waters, including shellfish waters, against pollution and, further, it is necessary to safeguard certain shellfish populations from various harmful consequences, resulting from the discharge of polluting substances into the sea. The Directive of 1979 therefore covers those coastal and brackish waters designated by the member states for protection or improvement to support shellfish life and growth. Shellfish culture is especially dependent on the maintenance of the high quality of coastal waters. Oysters, and sometimes mussels, are eaten raw, hence the preservation of public health requires close attention to water quality standards. The most important shellfish production basins need to be protected against hazardous effects of other human activities.

The approaches to regulating water quality are numerous. For example, there may be general prohibition (as in Spain), sanitary controls (Colombia and Mexico), general quality standards, sphere of protection (France), environmental impact studies (France), and elaborate approval systems (Colombia). All these laws provide for punishments ranging from fines to imprisonment, and from prohibition on sale to closing of aquaculture enterprises. One problem is whether enough information is available to assess the quality standards used, and the effectiveness of the methods to protect aquaculture operations. French legislation, for instance, provides for a variety of alternatives. It is possible to speculate whether aquaculture farmers really feel protected from large-scale pollution, such as

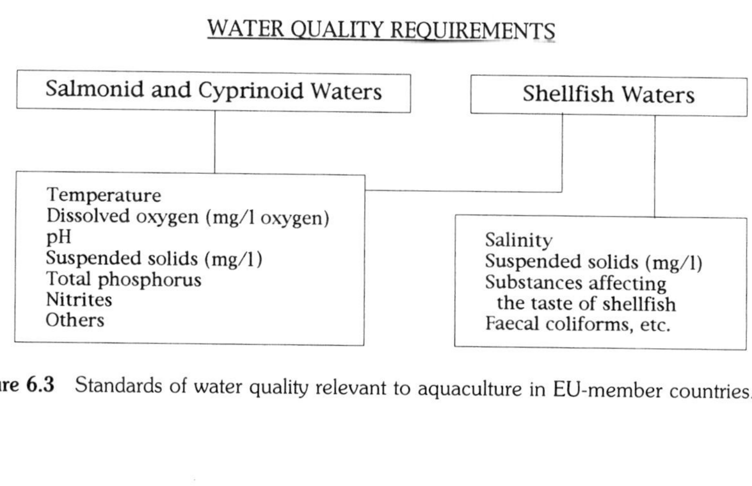

Figure 6.3 Standards of water quality relevant to aquaculture in EU-member countries.

industry, agriculture, and population centres, or if they themselves feel that they are polluters.

Finally, it is important to ask whether these various controls aimed at water quality, which operate at almost every phase of the aquaculture process, are more likely to lengthen procedures for obtaining the necessary approvals, or whether they do meet their intended objectives.

6.4 Environmental aspects

6.4.1 Aquaculture and the environment

Each country, whether industrially highly developed or less developed, has some level of environmental concern. Among environmental factors considered to have potential negative impacts on aquaculture development are pollution by domestic and industrial wastes, and toxic chemicals. In turn, aquaculture development itself may have adverse impacts on the natural environment, and on human health and welfare (see section 2.1). Many remedial and preventive approaches to integrate aquaculture development safely into the environment, and combinations thereof, are possible. Some of these are reinforced by legislation. Among many environmental issues relating to aquaculture practices and further development, the following principal issues have been identified which have legal implications:

- Land use planning and zoning,
- Control over installation and operation,
- Discharge limits and pre-discharge treatment to meet limits,
- Fiscal incentives,
- Prohibitions.

Land use planning and zoning

The recommendation of zoning areas for aquaculture purposes through legislative measures has been noted on water and land law; for example, in Ecuador, France, and Zambia (section 6.3). Such measures help in preventing the contamination of aquaculture products and in settling conflicts relating to land and water use, with the cooperation of other users in controlling environmental degradation and optimum use of land and water resources.

A number of countries provide for some type of protected areas for aquaculture. However, their regulations for these areas are all formulated differently, and modelled in different legal frameworks. For example, in Mexico the General Law on the Ecological Balance and Environment Protection provides for areas of protection in accordance with the Federal Hunting and Fishing Law, and exploitation and exploration of natural resources are subject to a licence. However, it is not clear whether aquaculture is included. In Malaysia the minister who has the responsibility for fisheries may, by order, establish any area or part of an area in Malaysian fisheries waters as a marine park or marine reserve; but in fact these are areas where aquaculture activities are not likely to take place because of all the restrictions. In the Philippines, certain projects, undertakings, and areas in the country have been declared as environmentally critical; no-one may undertake activities on or operate in any environmentally critical area without first securing an Environmental Compliance Certificate issued by the President or his duly authorized representative. Among projects declared to be environmentally critical are fisheries projects. In all these cases there are many restrictions, and much documentation is required, usually in the form of the environmental impact statement (EIS).

Control over installation and operation

Many countries require an environmental impact study to secure a licence for fish farming and to control any aquaculture enterprises proposed. The objective, as quoted in the regulations of the Philippines, is 'to attain and maintain a rational and orderly balance between socio-economic growth and environmental protection'. It can be assumed, therefore, that this broad legislation affects aquaculture operations, as areas declared as 'environmentally critical' are often suitable for fish farm installations and exploitation. Without due authorization issued by the respective ministry responsible for the environment and natural resources, no farming enterprise can be established. Such authorization is issued provided that agency approves the EIS submitted by the farming applicant. Guidelines regarding the contents and the procedure for such a study are usually provided by the ministry.

Typical requirements of an Environmental Impact Statement (EIS)

An EIS in Venezuela usually requires the following items:

- full description of the project,
- physical, biological, and socio-economic characteristics of the area of influence of the project,
- evaluation of the positive and negative effects on the environment,
- measures proposed to prevent, eliminate, or compensate for any harmful effects of the project on the environment, during the construction phase as well as the operative/active phase.

Source: Ministry of the Environment and Renewable Natural Resources, (Venezuela), undated.

At times the issue is taken further. In France, for example, the law adds that the content of the study is dependent on the importance of the planned work and on its foreseeable effects on the environment. Moreover, where an impact assessment is required there must be a public hearing.

Some countries also control operations by setting limits on activities by category, and consider permits or licences to prohibit what is not licensed, and to control what is licensed. In Norway, for example, the Ministry of the Environment has decided that fishing and fish farming facilities are regarded as ordinary pollution activities and that a permit must be applied for and considered in accordance with the Pollution Act. The State Pollution Control Authority evaluates all applications and grants permits to the Fisheries Authorities accordingly.

In summary, the targets of environmental legislation dealing with aquaculture are the natural resources to be protected against pollution or other deterioration, the considerations of effects on human health and welfare, the ecological impact, and the danger to undermine the regenerative capacity of the given ecosystem.

Discharge limits and pre-discharge treatment to meet limits

The implementation and/or adoption of environmental protection measures regarding waste discharge are common to many countries, and require various permits or licences, standards, pre-discharge treatments, and sometimes subsidies or charges/sanctions. For all intents and purposes, these are also included in the national requirements for the EIS, analysed in section 6.3.3.

Fiscal incentives

Fiscal incentives usually take the form of direct subsidies for the installation of the control equipment during a limited time period, or a tax deduction. All enterprises, to be granted the incentives, are usually classified and fulfil certain conditions, one of which is to install proper water discharge treatment. The incentives are exempt from duties normally payable on corporation or company operations, such as increase of capital, reconstruction, and share issues or division. Other exemptions include duties on working capital, export duties on fishery products, and import duties on goods and raw materials used in the processing of fishery products but not produced in the country, and the deduction from taxable income of up to a percentage (usually 50%) of income, loans or new capital reinvested in fleet or plant expansion or improvement. A more complete review of fiscal incentives used as policy instruments is given in Chapter 7.

Charges or taxes on polluting effluents or emissions resulting from aquaculture activities exist in Poland, Hungary, and France, amongst other countries.

Prohibitions of specific activities or definition of limits on activities

Finally, a number of countries prohibit specific activities or define certain limitations. In Norway, for example, a licence for breeding activities in freshwater, brackishwater, and salt water shall not be granted if the farm facility is badly located in relation to the surrounding environment, lawful traffic, or other on-going exploitation of the area. In the Philippines, measures for the rational exploitation of fisheries and other aquatic resources may include conserving the endangered species of fish and aquatic resources such as turtles, sea snakes, crocodiles, corals, as well as maintaining the mangrove areas, marshes and inland areas, coral reef areas, and islands serving as sanctuaries for fish and other aquatic life. In Ecuador, as in many other countries, destruction of mangroves is also strictly prohibited; and in Zambia the use of any pesticides as a means of curing, preserving, processing, or storing fish in any commercial fishing area is prohibited.

In summary, there are several existing environmental laws (and others) which relate to aquaculture activities. Some obviously discourage the establishment of aquaculture enterprises, and others reduce their effectiveness and efficiency, thus placing them in direct trade competition with producers in countries with less stringent regulations. However, these pollution-control approaches are part of a comprehensive international effort to use and manage resources properly. Pollution discharges can result in a body of water being useless for aquaculture purposes, in a way analogous to over-fishing. It is obvious that the further development of aquaculture in any country should be governed essentially by the available resources and the limits of environmental tolerance, in the local context. In the future all farmers will have to cope with and face other measures taken in view of environmental policies.

6.4.2 Fish diseases

Fish diseases can have dramatic consequences in any country with regard to its fish production and consumption, and even its foreign exchange earnings through the export of high-value products. The situation may be particularly consequential for individual fish farmers.

With regards to legislation on fish diseases, a difference exists between less developed countries and developing countries. In developing countries, such as Malaysia and Malawi, the competent minister is empowered through the national Fisheries Act to make regulations for this purpose, and/or control of fish disease is linked to rules dealing with fish import or export. Occasionally there is no specific reference to fish disease in the Act but, in these cases (as in Ecuador), it is common to find an Animal Health (Protection) Act which serves the same purposes, although it is often not always clear whether 'fish' is included in the definition of 'Animal'.

Developed countries, such as Norway, the UK, Canada, and New

Zealand, have legislation with preventive or remedial measures. These measures usually cover all aquaculture activities, whether in freshwater, brackishwater, and salt water. In Norway, for example, no licence to establish breeding activities shall be granted if the position of the site will cause risk of disease in indigenous fish and shellfish. In order to exercise some degree of control over the spread of fish disease from one farm to another the competent Minister of Agriculture, Fisheries and Food in the UK is given extensive powers under the Diseases of Fish Acts, 1937 and 1983, to take appropriate measures. Provisions cover both freshwater and marine fish. Provision is made for the registration of fish farms and the obligatory keeping of records of live fish and eggs transported both in and out of fish farms. Such records are designed to assist in tracing the whereabouts of suspect fish in the event of an outbreak of disease at their farm of origin.

Preventive actions or measures may also be vested in the licensee. In New Zealand, for example, the licensee, discovering or first suspecting any fish on the farm to be diseased, must advise the Director General for Fisheries or the authorized officer specified in the licence within 24 hours, and isolate them.

Fish diseases are usually controlled by preventive measures

Typical preventive measures used by countries are:

- different treatment and destruction actions depending on the kind of disease concerned,
- vaccination of freshwater fish provided that the vaccine is approved by the Minister,
- destruction of brood fish when genetic defects develop,
- sterilization of animal feeding material,
- prohibition to transfer diseased fish from any fish farm to any other fish farm or to release into any waters any diseased fish, and
- the issuing of certificates indicating the presence of any harmful or harmless disease at the moment of importation.

6.4.3 Importation of live fish and introduction of exotic species

Regulations on the import of live animals and fish are found in almost all countries. The scope of such regulations are either:

- the protection and control of animal health and prevention of animal disease generally, or
- the protection of aquaculture interests more specifically.

In most countries documentation is required prior to customs clearance at the border.

General rules to control animal health and disease

Regulations on import of live fish and introduction of exotic species, formulated to control animal health and prevent animal disease, are common to most countries. Such regulations may apply specifically to aquatic species suitable for aquaculture purposes, as found in Malaysia, Zambia, and the Philippines. According to their legislation no person shall import any live fish and/or introduce into any water of the country any non-indigenous fish without a written permit from the competent authorities. Terms and conditions of the permit often refer to the fishes' state of health, to avoid spread of disease, or to control release into the environment of non-indigenous species of fish. The majority of regulations, however, control entry into the country of all live animals, plants, and any biological products likely to carry any disease or disease agents.

Typical regulations regarding movement of aquaculture species are part of all general restrictions

'The import and export of animals (dead or alive), plants and items listed in the three Schedules to the Act are prohibited, except under a licence issued by the Secretary of State. All applications for licences are to be submitted by the Secretary of State to the appropriate scientific authority established or designated under the Act for its advice as to whether the licence should be issued. The licence issued under the Act may be general or specific, may be modified or revoked at any time by the Secretary of State and shall not in any case be valid for more than 12 months. The Act provides for the charging of fees, and the sanctioning of false or reckless application statements. In order to facilitate enforcement of the Act, the Secretary of State is empowered to make an order restricting the places at which live animals may be imported, where a licence has been given. Where a licence has been given to import a live animal of an endangered species covered by the Act, the Secretary of State may, where he considers it desirable, give a direction restricting the movement of the animal and requiring it to be held at specified premises, whether for quarantine or for other purposes.'

Source: Ministry of Agriculture, Fisheries, and Food, (UK), 1976.

Specific rules concerning fish disease prevention

The second category of rules on importation of live fish and exotic species is more oriented towards the protection of aquaculture interests and when importation is authorized mainly for aquaculture purposes. Such rules are applied in Venezuela, Panama, Mexico, Colombia, Spain, Canada, and France. Legislation is often quite recent (within the last 10 years, as in Venezuela, Mexico, and France) compared with others (some 20 years in Panama and Canada). Authorization usually takes the form of a permit.

General regulations for the authorization of importation of species for aquaculture

Regulations usually require the following information to obtain a permit:

- identity of applicant,
- scientific name of species (quantity and development status),
- purpose of import (aquaculture, scientific or ornamental purposes),
- source, place of origin,
- place of destination, and
- sanitary certificate from export country, especially for import of non-indigenous species.

A sanitary certificate (or certificate of health) is invariably required, which must state that the fish or eggs are free of any infections, diseases or parasites, or disease agents. In some countries, such as Mexico, Colombia and Spain, the applicant must provide, in addition to the items above, a technical study on the socio-economic impact and a description of the possible impact such species could have on the environment.

Furthermore, in some countries there are regulations or facilities for the quarantine of fish at introduction. Venezuela and Mexico, for example, impose quarantine restrictions to ensure disease-free importations and to observe the young resulting from controlled reproduction. Once this period of observation has passed there is often a final evaluation after which final authorization is granted or refused.

Some countries distinguish between importation of native and non-native live fish (e.g. France, Canada, and Spain). Others limit the authorization to a certain species (e.g. Venezuela to 'live shrimps', and Canada to cultured fish or eggs of wild fish of the salmonid family), or to live fish in general. Import of the latter usually requires, in addition to a certificate of health, a special consent. In France this is provided by the Ministry of Environment, and in Spain the Spanish Oceanographic Institute. Import of such species is often strictly limited for the purpose of fish culture.

In summary, regulations for the import of live fish seek either to control animal health or, more particularly, to develop aquaculture. These rules are usually enforced either by the withdrawal of the import licence or by payment of penalties.

Chapter 7
The Use of Policy Instruments II – Fiscal Measures and Support Services

Synopsis

This chapter introduces government policy instruments other than legislation, in particular the fiscal measures and supporting services which governments may or may not wish to use or provide to control and encourage development of the sector. It first describes the background of these measures. This is followed by the two main sections which separately examine the aims and objectives of fiscal measures and government support services to the aquaculture industry. These are illustrated by examples of policy instruments used by different governments in support of their sectors. However, these should not be taken as the ideal or, in fact, representative of an exhaustive review of the use of policy instruments in aquaculture around the world. The chapter ends with an assessment of these policy instruments and use to which they have been put.

7.1 Background

Present-day government policies and practices for developing and controlling the aquaculture sector have generally involved close relationships with other national sectors, particularly fisheries and agriculture. They have been used with the purpose of achieving certain political and social objectives in these areas, and to provide some degree of direction and control. Of particular importance to these three closely-related sectors is the way respective governments have tried to manage the utilization of key national resources, and allocate them fairly and in ways consistent with their policy objectives. This is achieved, as noted earlier, by the use of standard instruments of policy.

The most common and widely used instruments of policy are fiscal measures, of which probably the longest established is direct taxation. In raising taxes on aquaculture products governments have acquired resources from this emerging economic activity within their countries. Typically governments use varying rates of taxation to depress or stimulate economic activity, therefore, in the case of a new sector like aquaculture, taxation has been modest so far at best. Similarly, the ways governments then use public funds raised through taxation are also, in turn, further policy instruments. Again, as most aquaculture sectors are still only developing, a large amount of public funds has been used to manage the sector, to finance government support services for essential national and local infrastructure, such as research, education, and training, and provide incentives for producers, such as grants, credit, and subsidies.

Governments achieve policy objectives for aquaculture through groups of policy instruments

In addition to legislation, administrators and managers of national aquaculture sectors commonly use two other groups of policy instruments. These are:

- Fiscal measures, such as:
  - burdens (duties, taxes, and quotas),
  - incentives (grants, credit, and subsidies), and
- Support services, such as government services and physical infrastructure.

Fiscal measures

One group of fiscal measures are burdens on producers. These burdens are duties, taxes, and quotas, and include corporation taxes, income taxes, levies by government or semi-governmental agencies on sales of aquaculture products (including export sales), and quotas on levels of aquaculture production. The levels of these instruments applied to the producer by the national government affect the profitability of aquaculture production. Where products are exported the level of import duties applied by the government of the receiving country also has a significant effect on profitability.

The other group of measures are incentives for the sector through grants, credit, and subsidies, and include such things as grants for establishing or expanding aquaculture operations separately, or as part of regional development support, or to relieve the interest burden; equity participation by government agencies in aquaculture operations; lease of government-owned premises or facilities; government compensation and insurance schemes; credit for establishing or expanding aquaculture operations, or state guarantees for commercial loans; and subsidies for farm inputs such as feed, seed, fertilizers, power, labour costs, or price support subsidies, or construction by government of farms, hatcheries, and providing equipment at subsidized rates.

Government support services

As with all sectors, there are many services provided by government designed to support aquaculture. These include such things as market-related services, such as market information, or market intelligence, or information for consumers, or providing infrastructure for marketing aquaculture products; promotion of marketing organizations, or products, or technology; research and development in production technologies, or contract research; technical services for extension of technology, or education and training, or higher technical requirements (such as disease control and veterinary medicine), or business advice and support, or information; and general services for acquisition of external assistance, or collecting data and statistics on aquaculture production.

Often government support is active involvement in the sector and the provision of essential physical infrastructure. For aquaculture, typical infrastructure includes state farms, hatcheries, and other plant; transportation systems, such as piers, roads, canals, pumping stations; utilities, such as

Plate 7: A principal responsibility of the government is to collect and collate sectoral statistics. Technical staff in Dakar process data on the artisanal fisheries catch for member governments participating in a regional development project financed by the Danish International Development Agency and executed by FAO.
Photo: R. Cannarsa, Courtesy of FAO.

electrical power and water; and processing and distribution facilities, such as ice houses, refrigerated storage facilities, processing and depuration plants.

Government services which are institutional are part of the organization and management of the national sector. These have been described in some detail in Chapter 5.

7.2 Fiscal measures

7.2.1 Taxes, duties, and quotas

Manipulation of rates of taxation is a common way in which governments attempt to stimulate the process of economic development. This is true for both developed and developing economies. The application of favourable rates of taxation is aimed at stimulating investment, both from within the country and from foreign sources. For example, tax exemptions are commonly given on critical items of equipment, but this may differ between domestically manufactured and imported equipment. Exemptions are often offered on corporation taxes during the early stages of a project, and favourable rates of taxation are usually applied to earnings from export of aquaculture products in countries where the government is attempting to increase foreign exchange earnings.

A number of countries offer favourable rates of taxation on earnings for projects located in remote areas or regions where the government is giving special preference to economic development in association with social programmes. For example, government policy in Turkey is to encourage development of aquaculture production of high-value species for export. The 73 provinces have been grouped into one of the areas of priority in the Seventh Five-Year Plan, and those in the remote eastern region, the south-east, the north province of Zonguldak, and the Çorum-Çankiri region of central Anatolia, have been identified as the prime targets for development with financial incentives for national and foreign investment. The country has good potential for cultivation of marine species (sea bass, sea bream, and shrimp) along the coast, and freshwater species (trout, carps, eels, and crayfish) inland, and production has been established. Markets for products have been identified in southern Europe, and potential in the long term is considered good. The government offers a number of incentives to encourage development by attracting investment and technical knowledge.

Ecuador, on the other hand, has abundant natural resources for development of marine shrimp cultivation and now has a sector which has expanded rapidly and contributed to the country's export earnings for the last 10 years. Shrimp is now the country's second largest earner of foreign exchange, after petroleum. Several USA companies have invested and the government wishes to encourage further inward investment in the sector, particularly in hatcheries, to augment the supplies of post-larvae for the industry. To encourage further investment, the government has empowered the Inter-ministerial Committee of Industrial Development to decide on a series of benefits and incentives to offer to foreign companies seeking to invest in aquaculture activities in Ecuador, but the precise package of incentives is at the discretion of the Committee.

In Ireland, the rural areas of the west coast have been identified as being highly suitable for cultivation of salmon, mussels, and oysters. The area is considered to be disadvantaged socially and economically, and opportunities for creating employment and generating income are limited. For these reasons special emphasis is placed on development of aquacul-

---

Different countries stimulate investment by manipulating rates of taxation

Turkey
- total exemption from import duties on machinery and equipment,
- exemption from corporation tax until total initial investment is recovered,
- exemption from value added tax (VAT) on all exports,
- allowances of a percentage of all export earnings over a certain level deducted from total taxable income.

Ecuador
- full or partial exemption from import duty on equipment and raw materials,
- reduced taxation on fixed assets,
- greater incentives for projects located out of main urban centres of Guayaquil and Quito.

Ireland
- reduction in corporation tax for fish farms,
- tax relief on investments in fish farming on invested amounts up to a certain level.

Canada
- a machinery programme, where tax on imports of certain machinery not obtainable from Canadian manufacturers but important for advancing the industry is exempt from import tax for values above a certain level,
- exemptions of sales and excise tax for most goods for fish farmers (and fishermen) for their operations,
- fuel tax rebates for fish farmers (and fishermen) for diesel and gasoline used in commercial operations,
- provincial investment tax credit to a certain level for eligible costs of equipment purchased for operations in the Province of Cape Breton.

---

ture as a means of reversing declines in remote coastal communities. The technologies of salmon and mollusc cultivation are well established and industries are developing rapidly along the coast. The Irish government provides incentives in the form of tax exemptions to encourage further investment in the sector in west coast locations (see below).

Both the east and west coasts of Canada are highly suited to development of commercial aquaculture. The government is keen to promote the industry as a means of improving economic conditions in remote communities along these coasts. The country lacks key skills in the field of intensive aquaculture and is encouraging importation of technologies and equipment developed in industries of other countries. A range of tax exemptions have been agreed to encourage aquaculture development and transfer technology.

Import duties and quotas for fisheries products as a whole have been built up over the years to control trade between countries. Although not originally developed for aquaculture products most of these automatically affect the access of such products to foreign markets.

The most important international markets for high-value aquaculture production are the USA, Japan, and member countries of the EU. The following three examples illustrate the levels of duty to be paid for entry to these markets for (aquaculture) fisheries products from different countries, according to their needs and their own national or community production activities.

In Japan, for example (see Table 7.1), no duties are applied to fish fry (juveniles) imported for on-growing by producers in the aquaculture industry who need seed resources. Moderate levels of duty are applied to highly desired but unprocessed products, such as salmon (5%), and high levels applied to processed products, such as smoked salmon and shrimp, (15% and 7.5%, respectively).

The rates of import duty on aquaculture products applied by the EU demonstrate a variety of purposes to which this particular policy instrument is put (see Table 7.2). First, there is a high rate of duty (12%) on importation of trout. This can be interpreted as the EU protecting its own trout industries from cheaper imports from outside the EU. These trout industries, which are present in most member states, are certainly vulnerable to low price competition if duties of this scale are not applied. Import duties on fresh salmon, on the other hand, are negligible, and non-EU countries (such as Norway) can readily supply fresh salmon without suffering penalizing import costs. In this case EU member states are not protecting their own salmon-producing industries in contrast with its actions in relation to trout. The lack of any significant level of duty on fresh salmon imported into the EU also contrasts with the 13% levied on smoked salmon products. The effect of this action has been to make smoked salmon from Norway not competitive in the EU. Thus job opportunities in salmon-smoking in rural areas of Norway are lost, but employment in the EU's salmon-smoking industry, which is based mainly in France, is protected.

Table 7.1 Import duties on fish products from aquaculture for Japan

Product	Country category	Rate of duty (⇒ %)
Fry for fish culture	General	0.0
Live eel	General	10.0
	GATT	5.0
Salmon (fresh/frozen)	General	10.0
	GATT	5.0
Salmon (smoked)	General	15.0
Shrimp (fresh/frozen)	General	10.0
	GATT	3.1
Shrimp (smoked)	General	15.0
	GATT	7.5

Notes: (1) 'General' refers to countries with no trade agreements with Japan.
(2) 'GATT' refers to signatory countries General Agreements on Tariffs and Trade.
(3) Duty is calculated on CIF values of the imported products.
(4) Japan also applies annual quotas to imports of fish products These normally include products such as yellowtail, scallop, shellfish meats, kelp, and edible seaweeds.
(5) ⇒ means 'up to'.

Finally, the USA operates its own system of duties and quotas for countries which qualify for certain privileges of trade (see Table 7.3) based on national policies.

In summary, import duties are being used for a variety of purposes in relation to aquaculture. Japan 'zero rates' fish fry which are to be used for fish cultivation, where its own aquaculture industries depend on imports. The EU rates of duty create barriers to importation of trout to protect trout farming within the EU, and to the importation of smoked salmon to protect employment in salmon-smoking within the EU. In contrast high barriers are not provided to protect fresh salmon production in the EU.

Countries also impose limits on the amounts, or quotas, of fisheries products which can be imported and sold in their markets. All the main importing areas, namely North America, Japan, and EU countries, either apply quotas or have the legislation in place to permit them to do so when the need arises. The Japanese restrictions on imports of scallops is one

Table 7.2 Import dutices to the EU for fish products from aquaculture

Product	Country category	Rate of duty (⇒ %)
Trout (fresh/frozen)	General	16.0
	GATT	12.0
	Algeria	0.0
	Morocco	0.0
	Tunisia	0.0
	ACP, OCT	0.0
Salmon (fresh/frozen)	General	16.0
	GATT	2.0
	Algeria	0.0
	Morocco	0.0
	Tunisia	0.0
	ACP, OCT	0.0
Eels (fresh/frozen)	General	10.0
	GATT	3.0
	Algeria	0.0
	Morocco	0.0
	Tunisia	0.0
	Turkey	1.5
	ACP, OCT	0.0
Salmon (smoked)	General	16.0
	GATT	13.0
	Algeria	0.0
	Morocco	0.0
	Tunisia	0.0
	Turkey	5.2
	ACP, OCT	0.0

Notes: (1) The rates of import duty are imposed by the EU on aquaculture products from non-member Countries, or from specified groups of countries. No import duty is applied between EU members.

(2) ACP refers to African, Caribbean, and Pacific countries.

(3) OCT refers to overseas countries or territories of EU countries.

(4) EU also applies bilateral and global quotas on fish products but these do not currently affect aquaculture products.

(5) ⇒ means 'up to'.

Table 7.3 Import duties to the USA for typical aquaculture products

Product	Country category	Rate of duty (%)
Trout (fresh/frozen)	All	0.0
Salmon (fresh/frozen)	MFN	0.0
	GT	(2 cents/lb)
Salmon (salted)	MFN	3.1
	Special	3.0
	GT	25.0
	Caribbean	0.0
	Israel	0.0
Salmon (smoked)	MFN	5.0
	Special	0.0
	GT	25.0
Salmon (preserved/air tight)	MFN	12.5
	Caribbean	0.0
	Israel	10.0
	GT	30.0
Shrimp	All	0.0

Notes: (1) 'MFN' refers to countries with most favoured nation status.
(2) 'Special' refers to countries with a special agreement with the USA.
(3) 'GT' refers to all other countries.
(4) The USA does apply quotas to fisheries products from other countries but these currently do not affect aquaculture products.

example of quotas being used to protect a national aquaculture industry. The main purpose of quotas is to protect national industries from effects of market over-supply from imports. Other countries limit the quantity of imports of fisheries product as part of a more general effort to limit imports and protect the economy's balance of payments. However, some duties and quotas may have been sufficiently high or restrictive to make the difference between viability and non-viability for producers just starting aquaculture enterprises, as their products have immediately been susceptible to competition from outside the country.

7.2.2 Grants, credit, and subsidies

The common forms of financial incentives for aquaculture investments are grants, loans, and subsidies. Equity participation by governments in new high-risk ventures is also a further form of financial support.

The balance of use of these different types of financial assistance varies between the different types of economy. Subsidization of inputs, such as farm construction, public subsidy of feed, fertilizers, equipment, labour, and often prices, are more common and widespread in centrally planned economies and in developing countries, where the aquaculture sector is oriented towards subsistence farming or increasing local food supplies. The government of Nigeria, for example, includes subsidization of the costs of pond construction, seed supplies, and feed in its support for the aquaculture industry.

Subsidization, however, is considerably less common in countries with established aquaculture industries and particularly those geared to the production of high-value products. In these countries, and in developed economies generally, government grant and loan assistance schemes are more common. However, in common with most provisions for aquaculture support, these schemes are normally designed for wider application than aquaculture alone, with support for aquaculture included and made available through more general assistance schemes for agriculture and/or fisheries.

Among developed economies, the EU makes wide use of grants for aquaculture development (see Table 7.4). The levels of grant assistance vary according to the regional development status of the area with developments taking place. Within 'less favoured areas' higher rates of grants than in other parts of the Union are enjoyed. Capital costs of project development are usually the only costs eligible for assistance in grant schemes but, in some cases, grant aid may also be payable for operating costs. Currently, regions in the south of the EU area are attracting the higher rates of grant assistance.

EU grants are supplemented by further grant aid from member governments. In fact qualification for EU aid is conditional on such aid. The EU

Table 7.4 Grant assistance for aquaculture in EU countries

Regions designated for assistance	Maximum grant
Less favoured areas: (Greece, Andalusia, Canaries, Castile-Leon, Castile-La Mancha, Galicia, West of Scotland, Ireland, Northern Ireland, Mezzogiorno (Italy), Portugal, French Overseas Departments).	40%
Other EU areas	25%

member governments in their turn have different levels and conditions for grants. For example, financial assistance from Denmark is in the form of a 10% grant for construction, modernization, or extension of aquaculture production units. The grant rate is increased up to 25% if the project includes provision for protection of the environment. Greece, on the other hand, provides 20% of capital costs for installation and equipment for farms. Up to 50% grants from national sources are also available together with subsidized interest on loans required for investment in aquaculture. Up to 15% additional grants are available to local government organizations and Greek nationals returning from abroad, and for Greek merchant seamen. In Ireland there are 50% grants for pilot project capital costs up to a maximum level, and 10% grants for capital costs of commercial projects. There are also loans for bridging purposes of up to 20% of project costs. The Netherlands provides 10% grants on project capital costs, and financial guarantees for commercial loans up to a maximum level. Finally, Spain provides 30% grants for projects located in 'less favoured areas', as defined by the EU, and 10% for projects in other areas. It also provides financial guarantees for projects submitted to the EU for funding. Assistance is also provided in the form of subsidized interest on market loans.

Many countries provide grants to reduce the costs of lending from commercial sources. This are *interest relief grants* or *subsidized credit schemes*. For example, this type of grant is available to aquaculture producers from the UK's Highlands and Islands Development Board (HIDB). Another scheme is for governments to reduce unemployment through the subsidization of farm labour costs.

Variation in the rates of grant assistance and subsidized credit is another way in which governments seek to achieve policy aims. The Danish government, for example, provides higher rates of grants for projects which it considers to be environmentally safe. Canada gives special rates to projects sponsored by members of native populations, Ireland to speakers of the indigenous language, and in Greece redundant merchant seamen and returning exiles can receive special assistance. The most common reason for higher rates of assistance, however, is to direct development to particular areas within a country. Aquaculture projects in the Western Isles region of the UK – an economically difficult area – attract very high rates of grants from the UK government and the EU.

The most important way governments support the development of the aquaculture sector is through the provision of credit. Like grants, credit is provided through a variety of means.

Some governments, through national banks or quasi-government organizations, provide financial credit by taking *equity shares* in companies involved in innovative or high-risk development projects. For example, the UK, through its HIDB, held equity in commercial salmon farms during the early phases of that industry's development in Scotland; the Irish government has acted similarly to support shellfish cultivation, and Canada's Federal Business Development Bank Venture Capital scheme involves temporary shareholding. France provides joint-venture credit for national investors overseas through the Caisse Central.

Credit schemes are administered in a variety of ways. These include direct administration by the government ministry or indirect administration through the private sector, such as commercial banks, or through development authorities. The Irish Fisheries Development Board (Bord Iascaigh na Mhara) and the HIDB are examples of development authorities which have a more general remit, but which have invested in aquaculture to significant levels. Assistance schemes are commonly backed up by provisions for leasing land and facilities from government.

Specialized credit schemes for aquaculture are rare. Most farmers obtaining credit either from commercial banks or, more commonly, from credit schemes set up by government for agriculture generally or for small business development. In the USA and Canada, and in Europe, the more common sources of credit for aquaculture have been government supported loan schemes. In the USA, for example, these have included the Farmers Home Administration, the Farm Credit Association, and the Small Business Administration.

In the developing economies there have been major problems in the provision of credit for economic activities in rural areas and, where aquaculture is a new local activity, establishing credit from existing sources has been particularly difficult. There have been some successes. In India and Bangladesh, for example, there have been active programmes established by rural development banks. Even in these cases, however, many small-scale farmers continue to prefer to use local money-lenders.

Most governments operate credit schemes for small-scale farmers

In 1977 the government of Bangladesh established a special credit scheme with a component for fish farmers. The terms of the loans are for five years at 16% annual interest rate but with a deferred first payment for 18 months.

The normal channels for the provision of credit for aquaculture in developed countries are through government schemes for the development and support of agriculture or small businesses. Specialized schemes of credit for aquaculture alone are the exception rather than the rule and, as in developing countries, the commercial banking sector's credit terms have prevented all but the large projects undertaken by major companies being supported by investment loans or credit from this source.

The governments of most developed countries provide a range of agriculture and business support schemes which contain loan assistance as an element. These are instruments designed to implement government policies, such as protection of living standards in the agricultural sector, strengthening of rural economies, creation of job opportunities in rural areas, and protection of the small-scale business sector.

The provisions of the federal and provincial governments of Canada illustrate the wide range of credit schemes available from such sources. Between them, they offer 16 different schemes for private enterprises in the provinces of Canada. Some illustrate the fact that many of the schemes from which aquaculture can obtain financial support are not specifically targeted toward aquaculture but are designed for more general regional development; for example, the Canada–British Columbia Subsidiary Agreement on Small Business Incentives to encourage the establishment, expansion, or modernization of small and medium-sized businesses in British Columbia; or the Special Agricultural and Rural Development Act aimed at economic improvement of residents of remote and northern areas of Canada of native ancestry. Another loan scheme has been introduced to compensate mollusc growers whose markets were temporarily ruined by a ban on sales following a shellfish health scare.

In developing countries commercial banks have credit facilities for farmers but they normally only lend amounts which are usually too large for small-scale rural farmers. Commercial banks require collateral for loans, and these have to be in the form of land title or other company investments. Frequently fish farmers do not have proper legal title to their land, while the lease of a pond or pond complex, or the value of its stock, are often unacceptable or insufficient as collateral. These commercial banks, therefore, have not generally been a source of credit for aquaculture development.

The development and maintenance of a new aquaculture sector at the small-scale, artisanal level therefore requires provision of credit from other sources, such as cooperative-based institutions, government agricultural credit programmes, or state-owned or controlled banks.

Clearly, the problem of securing and designing successful credit for commercial aquaculture in developing countries is complex, and involves difficult choices. There are considerations of the degree of professional management, lending institutions' technical knowledge in the field of aquaculture, the ability of institutions and credit arrangements to maintain themselves through recycling of repaid loans, the varying abilities of institutions to operate in rural areas, and the degree of complexity of the procedures for acquiring loans.

Agricultural credit from non-commercial sources has also been available from public parastatal institutions. In Zambia, for example, there is the Agricultural Finance Company, the Zambia Agricultural Development Bank, the Development Bank of Zambia, and various cooperative financing institutions. There has, however, been no substantial provision of credit for small-scale fish farming. The sector is new and, although good commercial prospects have been identified by pilot projects, aquaculture does not have a sufficiently long track record for there to be adequate expertise in funding the activity. The only credit schemes for aquaculture have been funded by international agencies, and have been experimental in nature.

Even where technical and economic viability of farming options has been confirmed, the successful establishment of commercial farms depends on the availability of funds for the sector. This, in turn, depends on the levels of awareness of the true potential and needs of commercial aquaculture in the country by the banks and other lending institutions. It also depends on how strong competing demands on funds are from other sources within the economy.

7.3 Government support services

7.3.1 Development of markets for aquaculture products

Increases in product supplies from aquaculture invariably compel needs to increase market development and support. It is normal for governments to provide support services in this area, although centrally planned and other economies differ mainly in respect in the degree to which the public sector intervenes.

These services are usually provided through organizations which take the form of a parastatal authority. They may be created entirely with state support, or with joint contributions from the state and the industry itself. The placing of these authorities within the national organizational structure, and their typical objectives and functions, are discussed and illustrated by examples in Chapter 5 on organization and management. In some cases the authority is concerned only with marketing (such as the Fish Marketing Organization of Thailand) and in others marketing may be only one of its functions (such as the New Zealand Fishing Industry Board, or the Sea Fisheries Industries Authority of the UK).

Market services for fisheries products are generally more highly developed by comparison with other aquaculture support services, given the longer history of government involvement in the support and control of markets for products of capture fisheries. Some services include provision of market infrastructure, possibly transportation links and market structures, but most services are confined to marketing information and intelligence. This includes collection, collation, and publication (through a variety of media) of current FOB and wholesale prices, product availability, and holdings. A more interventionist policy involves public sector selling within markets where prices are subject to tighter government control, but this is not common.

In addition to information for the trade, some services are directed towards the consumer. Again these may include current retail prices, and availability, and also consumer advice.

Closely associated with market-related services for both the producer and the consumer are promotions. These may be continuing services, such as the promotion of a marketing organization within the industry, or services which include brief finite promotions, such as campaigns to promote aquaculture products as a whole, or specific species, such as salmon, trout, or oysters. For example, the French government provides state support for the promotion of farmed trout through advertising and market promotions jointly funded by government and the industry.

In certain cases, if the aquaculture products are sold on international markets, governments may establish an international sales organization, but usually they directly assist sales and marketing associations established by producers. Such associations promote the sale of the products through international advertising, underwrite representation at sales fairs and exhibitions, and may underwrite some of the transportation costs through the national airline.

Governments support sales and marketing associations formed by the producers themselves

In Norway the Norwegian Fish Farmers Association (Norske Fiske-oppdretternes Forening) is a nationwide organization for all fish, smolt, and shellfish producers. The Association functions as the voice of the industry and negotiates with government authorities regarding the framework of the industry. The Fish Farmers Sales Organization (Fiskeoppdretternes Salgslag) is owned by the fish farmers and controls the first-hand sales of all farmed fish. It approves buyers, who take care of distribution of sales, including exports, after negotiation of minimum prices. It also assists with the buyers in marketing. It charges both sellers and buyers to finance its activities. At international exhibitions and fairs the group, and other national companies, are assisted by the government with transportation, exhibition space, and other activities, to promote international markets. The two organizations also own and operate the Norwegian Fish Farmers Breeding Centre, and produce a journal for farmers, *Norsk Fiskeoppdrett*.
Note: In 1991 the Sales Organization terminated its operations due to financial reasons and inferences that it was over-reaching its legal mandate. Nothing has replaced it as yet, due to the continuing litigation process. In all probability, if Norway becomes a member of the EU, a Production Organization will be formed, similar to others of the EU which are not under the control of producers. The Farmers Association continues to operate.

Other promotions which might be handled by this or another authority include the development and promotion of new product technologies. These may be quite elaborate, particularly if the authority was an investor in the development process, or again brief and finite. For example, like the example noted above, at international trade shows all manufacturers or service suppliers from one country may exhibit under one national display co-sponsored by the authority, or another government department, such as trade and industry.

Many producers find advantages through the formation of their own service organizations, independent of government. These may be small associations, national federations, or cooperatives. The advantages vary. They may be financial, through the collective purchasing of supplies, such as seed, feed, and fertilizer, or through collective distribution and handling of the products, including marketing under a common brand name; they may be political, through lobbying for government programmes, such as subsidies on feed or investment in research; or they may be simply useful, through the publication and exchange of technical information, prices, and marketing information. Another important advantage of such organizations is that the quality of the product is substantially upgraded as the producers all accept higher standards for harvesting and post-harvest handling.

Such organizations are established independently by the producers themselves. Similar organizations are often established by feed manufacturers and other suppliers, and may be international in scope; for example, the Fédération Européenne de la Salmoniculture (FES) is the central international organization which includes all the national salmonid (salmon and trout) culturists' federations and associations of the European countries. It assists its members in dealing with differences in national legislation, and in regional trade of salmonids, in which the trade in live fish is an essential component. In 1994, the FES opened its doors to all producers of fish and has been renamed the Federation of European Aquaculture Producers (FEAP).

7.3.2 Develop research, and provide education and training

Research services, especially in applied research, are needed by aquaculture industries. These are normally undertaken by national governments or by international organizations working in conjunction with governments. The introduction of new technologies, or the adaptation of existing ones to local conditions, depends on programmes of research. In most countries these tasks are undertaken by public agencies. There are, of course, exceptions to this. For example, several private companies have developed intensive production systems for high-value products based on their own research efforts; and many of the commercially-produced fish feeds are based on formulations developed through private research. Private companies are also active in research developing veterinary medicines.

Aquaculture research is normally carried out at universities, marine science institutes, or laboratories. Institutions specifically geared to aquaculture research are relatively rare, but do exist.

Aquaculture research capabilities vary considerably from country to country, and by region. The strength of the capabilities reflects the government's degree of commitment to aquaculture development and its financial

Plate 8: The need for education and technical training services is met largely by the government. Students at the Freshwater Fisheries Research Centre at Wuxi, China, are receiving practical instruction in the breeding of carp. The Centre is operated by the Chinese Academy of Fisheries Science.

support. At the present time North America, Europe, and the aquaculturally-developed countries of Asia dominate research outputs. Many of the aquaculturally-developed countries also export research services to developing countries.

Research communications with farmers are frequently complex

In Indonesia three Ministries (Agriculture, Education and Culture, Population and Environment) support research, education, and training in aquaculture. The Ministry of Agriculture provides support services through (a) the Agency for Agricultural Education, Training and Extension, and (b) the Agency for Agricultural Research and Development. The services are delivered through a Coastal Aquaculture Research Institute, a Brackishwater Training Centre, two High Schools, a Fisheries Academy, and a range of short-term courses in different parts of the country. The Ministry of Education and Culture provides support for education, research, and extension through its Department of Higher Education. Services are delivered at ten universities in the country. The Ministry of Population and Environment also provides support for education and research at a number of environmental study centres at universities throughout the country.

The nature of responsibility for aquaculture research in Indonesia, for example, and how its transfer to farmers is organized at the national level, is often complex. The arrangements involve considerable overlap of functions between three ministries. In this respect, and in the relatively dominant role of the Ministry of Agriculture, the arrangements for support services are typical of many other countries.

Complexity is not always the case. In Taiwan, for example, the Ministry of Education provides funds for research and education at the National Taiwan University, the National College of Marine Science and Technology, the National Pingtung Institute of Agriculture, the Bureau of Fisheries, the Fisheries Research institute, and several provincial marine high schools. Although these many establishments deliver support services to the sector they are all under the control of one government organization.

Common problems with aquaculture research programmes include lack of applied research and over-emphasis on biological aspects of aquaculture. In many countries, particularly in Africa, research programmes are severely constrained by lack of funds, poorly maintained facilities, and inadequate equipment.

The need for *education and training services* in aquaculture is also met largely by government, although some private companies in developed economies provide short-term training in practical and management skills. The education and training needs of aquaculture are wide and at a number of levels (see Figure 7.1).

Training provision should be based on realistic manpower demand assessments. These assessments should consider both the numbers of trained personnel required and the levels and specialized areas of skill which countries' aquaculture industries require (see Table 7.5). The development of aquaculture in many countries is constrained by the lack of critical skills and the lack of provision of training to produce these skills. On the other hand, some governments, following over-optimistic assessments of demand, make an over-provision of training with the result that the number of trainees exceeds the demand in existing industries.

Where aquaculture is well established there is normally a wide range of training facilities and programmes at the different levels. Common problems with training include over-specialization in certain aspects of

The nature and level of training can be categorized for aquaculture in the following way
• Training for practising fish farmers • Training for fish farm technicians • Training for government workers • Training for trainers

Figure 7.1 Government training services.

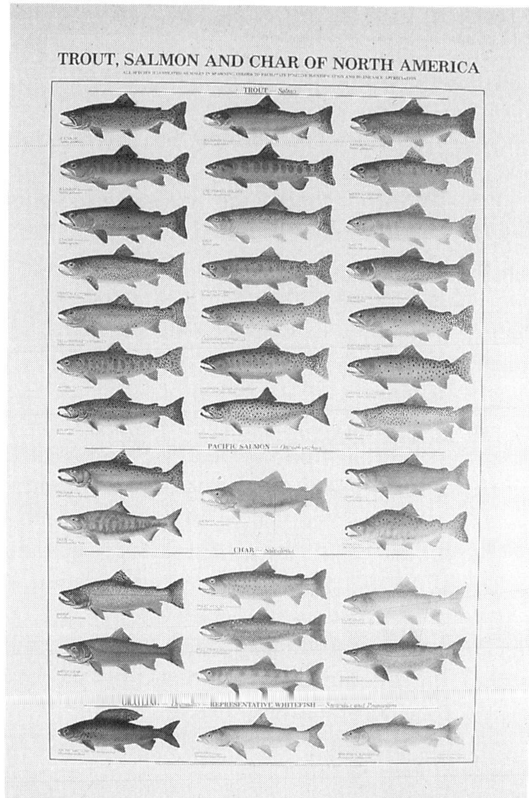

Plate 9: Additional government responsibilities for sector development include information and workers' safety standards. Plate 9a (top) illustrates a typical popular wall-poster published by many government departments and fishermen's associations for public awareness about aquaculture and seafood products: Plate 9b (bottom) shows a Japanese workman suitably clothed and operating safely on floating cages.
Photos: Courtesy of Conrad V. W. Mahnken.

Table 7.5 Assessments of training requirements and skills are important for planning (number of personnel required)

Country	Planning/ management	Hatchery technicians	Farm Systems (a)	Technology (b)
Bangladesh	29	49	63	32
Burma	26	42	21	26
Brunei	5	8	13	6
China	60	10	17	75
Hong Kong	1	3	2	4
India	135	82	110	134
Indonesia	59	79	208	92
Rep. of Korea	11	5	11	8
Malaysia	54	51	52	49
Nepal	30	18	24	22
Pakistan	34	26	27	38
Philippines	92	39	75	68
Singapore	3	1	–	4
Sri Lanka	32	13	25	18
Thailand	32	16	31	39

Source: Chua, 1986.

Notes: (a) Systems include integrated fish farming, brackishwater fish culture in ponds, cage and pen culture, mollusc culture.

(b) Specializations include aquaculture engineering, fish diseases, fish nutrition, feed technology, and genetics.

aquaculture and lack of practical content related to the production requirements of the industry. Also, training provisions are often based on inaccurate assessment of industries' manpower requirements with the result that there is under- or over-supply of some skills. Training facilities in many of the developing countries are also often of a poor standard and equipment for training inadequate or out of date.

In almost every country there is a range of institutions providing fisheries and aquaculture training

Typical institutions providing aquaculture training are:

- educational institutions (schools, colleges, universities),
- government institutions (national, regional and local government departments involved in aquaculture development or control),
- bilateral and multilateral agencies and international organizations.

7.3.3 Increase production by providing technically-related services

Technically-related services are provided at a number of levels in the sector, but are more extensive and greater in support of producers. These services are centred around on-growing, harvesting, and post-harvest technology.

In centrally planned economies these support services are not as distinct as in free-market economies, as production and post-harvest handling are functions already undertaken by state-controlled agencies. In Hungary, for example, until recently production and processing were mainly carried out by state farms and cooperatives, and these were ultimately controlled and serviced by the Ministry of Agriculture. In Cuba, the National Aquaculture Enterprise undertakes production directly in addition to other service functions, and is also under the control of the Ministry of Agriculture. However, these state agencies differ from government departments in that they have state enterprise status and, as such, have some degree of independence of action from government departments.

In free-market economies direct state intervention in support of private enterprise production is largely avoided because it is less cost-effective and not as reliable as necessary. However, there are some exceptions which mainly concern the enhancement of public waters (see section 1.3.2). Public hatcheries are used to produce juveniles for the culture-based fisheries (such as the salmon fisheries of Japan and the USA), and sports fisheries. However, intervention is mostly confined to hatchery production, and the supply of seed for farm production and culture-based fisheries is a part of the support services for local infrastructure which are dealt with in the following section.

Support services which provide very specific direct inputs to the producers collectively form the 'local infrastructure' of the sector. There are many potential services for governments to undertake, among which are the following.

The supply of seed

All aquaculture, whether for aquatic farming or for enhancement of fisheries, is totally dependent on the annual supply of seed. Seed describes fry or fingerlings in the case of fish species, post-larvae in the case of crustaceans (shrimps and prawns), and spat in the case of molluscs. It is also important for farmers that supplies of seed are reliable each year, that the numbers required are delivered on time, the quality and viability are high, and the price is competitive.

Many farmers, and not always those operating large vertically integrated farms, produce their own seed for the reasons outlined above, although this is not always economical. Most small-scale farmers are compelled to buy their seed, unless the propagation techniques are simple, or well integrated into the natural spawning seasons, as is the case of oyster and mussel farmers placing seed collectors in the water.

For some species of fish, such as milkfish, mullets, and eels, where the propagation technology is not yet developed sufficiently to be reliable but grow-out is successful and profitable, farmers buy their seed directly from collectors of wild resources. In some countries, such as Taiwan and Egypt, the government intervenes to control prices and distribution of these resources, and maintain some quality levels for the farmers as many of the wild resources are mishandled and moribund. They have established government centres which act as focal points for collectors and farmers alike. In others countries, such as the Philippines, collection and distribution are left to free enterprise, but the government maintains control by licensing the middlemen or brokers.

Governments often intervene in farming practices still dependent on wild resources of seed

In Taiwan all milkfish seed are distributed through six government centres, where they are counted and sold to farmers. In Egypt collection and distribution of mullet fry is controlled by the government and some 200 million seed are handled annually by the government farms. In the Philippines the collecting industry for milkfish is an important economic activity for many small coastal communities and licensed distributors. Together they collect and distribute some 300 million milkfish seed annually direct to farmers throughout the country.

For marine shrimp seed, there are frequently options to buy both wild seed (post-larvae) and propagated seed from hatcheries. This has invariably caused instability in the industry.

Alternative sources of seed can cause instability in the sector

In Ecuador farmers prefer wild resources of post-larvae of marine shrimps as they are cheaper. Unfortunately wild resources are not always reliable each year and farmers are then prepared to buy propagated stocks from private and publicly-backed hatcheries. The hatcheries, however, are not always ready to meet the shortfall as it is not economical to operate highly technical and labour intensive hatcheries on speculation. Consequently in some years national production drops because of the unnecessary shortage of seed resources.

Reliable annual sources of seed are fundamental to the industry, and in most cases the propagation of seed rapidly becomes an economic sub-industry of its own. For example, in Sweden there are 170 private salmon smolt producers supplying about 250 farm sites; in Thailand there are over 1000 small backyard hatcheries supplying post-larvae freshwater prawns and marine shrimp for growers. This rapid growth in private hatcheries is characteristic of national industries based on high-value species, such as Atlantic and Pacific salmon, marine shrimps, and freshwater prawns. For these species, the margins of profit are large enough for both farmers and hatchery producers to make money.

In conclusion, there is little or no role for government to play in the peripheral sub-industry of seed supply for the high-value species. The private sector proves that it responds efficiently and cost-effectively to the needs of the farmers, and in both developed and developing countries. There are some exceptions for efficient government intervention, namely:

- countries with centrally-planned economies to operate state hatcheries to supply cooperatives and state farms with seed, and
- federal and state governments of any country may invest in the construction and operation of hatcheries to support the introduction of aquaculture.

Public investment in demonstration hatcheries encourages investment

In the State of Hawaii the Department of Land and Natural Resources supplied farmers with post-larvae of freshwater prawns for many years until the industry was established; the hatchery was then sold to private enterprise. The Governments of Côte d'Ivoire and the Central African Republic each built and operated, with technical assistance, an African catfish hatchery to provide seed for farmers to encourage investment in the industry.

The most appropriate role for government in hatchery propagation is in support of enhancement, particularly for culture-based fisheries and recreational fisheries. The third option of enhancement, namely ranching, is not appropriate for government investment because of the high risk and private nature (ownership of released stocks) of this activity.

Culture-based fisheries are in the public domain, even though there are direct financial benefits for commercial and small-scale fishermen. A fishing industry is important to governments in that it fulfils political, economic, and social objectives, and has both direct and indirect returns on government investment. The benefits are those typical of any sector, namely contributions to food self-sufficiency, employment, and localized economies. Because of the revenues which accrue to the government in the form of taxes and licences, governments are justified in their investment in constructing and operating propagation hatcheries.

Enhancement programmes may involve federal, state, and private hatcheries

In Japan over half of the Pacific salmon taken in the high-seas fishery are propagated and released from hatcheries. The Island of Hokkaido has 65 hatcheries on its waterways operated publicly and privately but under the control of the Ministry of Fisheries. In addition some 20 million fish return to be taken in licensed traps by coastal fishermen. Japan's enhancement programme releases over 2.5 thousand million seedstock annually, including salmon, red sea bream, kuruma prawn, and scallops. In the USA, the Columbia River System has about 100 hatcheries on its waterways propagating and releasing Pacific salmon species. These hatcheries are owned and operated by the federal government, three state governments, private utility authorities, and American Indian tribal communities.

Culture-based fisheries include also compensatory programmes in response to national investment in irrigation and hydro-electric schemes which destroy natural resources of migratory fish. Hatchery propagation for compensatory stocks are therefore the responsibility of the government and/or their parastatal authorities which operate them.

By law governments may be required to operate compensation programmes of enhancement

- In the USA, the Columbia River Project, which has been a 50-year effort to bring irrigation water from three major rivers, has over 30 hatcheries for Pacific salmon operated by the federal government built to compensate for the loss of breeding grounds and blockage of migratory waterways with dams for reservoirs and hydro-electric power generation.
- The Baltic countries collectively release almost 6 million salmon smolts into the watersheds to compensate for the loss of the offshore fisheries through hydropower schemes.

Finally, recreational fisheries are also almost entirely in the public domain and are therefore the responsibility of government, rather than the private sector. They fulfil social objectives and programmes as they contribute to the general quality of life. They also have direct financial returns on government investment because of the revenues which accrue in the form of taxes and licences, and indirect returns through the localized economies. Governments are therefore justified in their investment in recreational fisheries programmes, and many federal and state governments construct and operate hatcheries for producing seed to stock recreational fisheries.

Many governments support recreational fisheries

- In Hungary the hatcheries of state farms and cooperatives seed 27 000 ha of waters with about 12 species of fish for some 30 000 registered anglers.
- In the USA almost every one of the 52 state governments has a fisheries and wildlife department which operates an enhancement programme for recreational fishermen. Facilities are not small. A warmwater fish hatchery in the State of Kansas produces annually some 2 million catfish, 400 000 largemouth bass, and up to 40 million walleye fry.

Supply feed and fertilizers

After seed, feed is the priority for the producers. The majority of farmers who produce high-value fish and crustaceans, most of which are carnivorous or omnivorous species, supplement the natural productivity with prepared artificial feeds, or alternatively provide all the nutritional requirements through such feeds.

Artificial feeds have to be manufactured to specification for both size of pellets and nutritional quality. Different species often require different feed ingredients or different percentage compositions, and there may be differences within species for broodstocks, seedstocks, and grow-out stocks. A typical feed is made up of 20–40% animal protein. As this is invariably the most expensive ingredient it is kept as low as possible, except for broodstocks and seedstocks. These require high quality feeds which can be affordable as the volumes required are relatively small. For the large quantities of feed required for grow out, the animal protein composition is minimized. The rest of the feed is made up of plant proteins, carbohydrates, fats, trace elements, and vitamin supplements, together with water. The feed may be dry or moist, depending on the percentage of water.

The cost of artificial feeds is important to the economics of most farm operations

Feed costs are the most significant factor in the operational costs of most types of farms, and may be as high as 60–75%. Depending on many factors, the conversion rate of a kilogram of artificial feed to a kilogram of final product varies between about 1.5 and 5:1. With poor husbandry, overfeeding, and losses, this can increase to uneconomic proportions. Consequently, in addition to the manufacture of feed, producers require a feed manufacturing industry which will produce and distribute the feed properly, and provide professional advice on storage on the farm, feed utilization, and feeding equipment.

There is no special role for government in services related to the supply of feed for producers. Artificial feeds for aquatic animals, for all intents and purposes, are similar to those for land animals and poultry, and most manufacturers can be persuaded to produce to specification, batch lots of feeds for fish and crustaceans. Most countries have either private or government-owned feed manufacturers which can meet the demands of producers. The requirements of government are therefore to encourage participation by the private manufacturers, or to produce feed if the industry is government-owned.

On the other hand, some countries which have limited resources for ingredients and require costly imports (particularly island countries, such as Malta), subsidize the cost of animal feeds, or exercise price controls. These may or may not include feeds for aquaculture.

For many farmers who produce low-value species of freshwater fish, most of which feed on naturally occurring aquatic plants and planktonic organisms, the availability of food is achieved by good pond and water management, and good husbandry. The natural productivity of ponds can be increased by the addition of composts and organic wastes, some of which also feed the fish directly, and also by the careful addition of inorganic fertilizers, and also the productivity of coastal waters for the increased production of molluscs, such as oysters and mussels.

Fertilizers are often used in aquaculture, either broadcast over the water, or allowed to leach into the water from release points, or ploughed into the substrate when the ponds are dry. There are no specific criteria as yet for fertilizers for aquatic animal production, and standard agricultural fertilizers are used, produced by private or government-owned manufacturers. The role of government in meeting the needs of producers for fertilizers is the same as that for feeds.

As noted above, several governments subsidize the costs of feeds and fertilizers for the agricultural sector, and consequently some of these same governments apply the same subsidies for the aquaculture sector. Others offer farmers tax relief on their feed and fertilizer costs.

Supply equipment

Producers require a considerable range of equipment, depending on the level of production technology characteristic of the country concerned. Much necessary equipment is readily obtainable from other industries, such as pumps, valves, and pipework, or can be fabricated or adapted from other industries, such as nets, tanks, boats, and processing machines. Most laboratory equipment is available, such as ultra-violet sterilizers and water quality instruments, together with necessary chemicals and veterinary medicines. Some specialized equipment is necessary, such as floating cages, automatic feeders, and graders.

For the most part the private sector active in these areas responds to the opportunities of a new market for their products, without the need for government intervention. These opportunities are made more apparent through trade shows and exhibitions.

In some countries, however, equipment is not readily available and has to be imported. This is a constraint to producers in developing countries where there is a shortage of foreign currency. Governments can assist such producers through policy instruments, for example, duty free importation, subsidies, and tax exemptions, which may be applied to both agriculture and fisheries sectors (section 7.2).

Governments assist in the purchase of operating equipment

In Saudi Arabia aquaculture producers qualify for import duty relief on pumps and pipework as irrigation equipment through the agriculture sector; and nets for floating cages as fishing gear through the fisheries sector. The existence of a broad range of equipment suppliers and services to the producers should not be underestimated by governments. In a survey of the industry in the UK, which produced about 25 000 t per annum, valued at £65 million (1986 data), it was estimated that the market for goods and services could be worth £1900 million a year between 1988 and 1992, with £200 million for capital equipment, and £1700 million for services such as the supply of feeds.

Provide extension services

The extension of technical information is a continuous need of any aquaculture sector, whether it is new or whether it is long established. Extension is predominantly considered a responsibility of governments, both federal and state, and many countries are establishing extension services for aquaculture either individually, or in association with extension programmes for agriculture, small-scale fisheries, and even wildlife management. Extension may also be handled by the private sector. This is achieved through private consultants, and by the companies which manufacture feeds, veterinary medicines, and equipment, etc., which send representatives to instruct farmers on the use of their supplies, and who leave printed information.

Because the resources for aquaculture, the production practices which are developed, and the farmers themselves are 'local' in nature, extension itself is a 'local' activity. Extension agents, in addition to being technically trained, must know characteristics of the region and, if necessary, communicate with the farmers in their local language or dialect. Thus, in addition to frequent visits by the agents to the farmers, the farmers themselves should be able to visit local extension offices for information, or government facilities for demonstration courses and training.

Extension services can be large and costly to operate. In addition to construction and maintenance of facilities, there are salaries and wages of full-time trained personnel, and requirements for extension materials, transport, transportation expenses, and instructional courses.

Extension services may be large and costly to operate

In the Philippines extension is executed by the Bureau for Fisheries and Aquatic Resources through its 12 regional offices, and provincial and municipal offices. There are some 300 extension officers, representing about 50% of the department's force, with each assigned to cover an area of about 120 ha. In support of the programme, the department has 31 freshwater farms, 32 brackish-water farms, and 11 marine farms for seed production, field testing, and demonstration of aquaculture technologies. The Philippines produces over 700 000 t of mixed aquaculture products annually, valued at US$1.0 billion (in 1992). Extension officers are among the most lowly paid of government workers, and the service is usually poorly equipped. Responsibility for servicing the profitable marine shrimp industry in the Philippines has been taken over by the private sector, including the large vertically integrated farm enterprises which include consulting services among their profit centres.

Because of the logistics of extension most governments adopt a central organization for all their extension requirements, together with activities beyond production. Israel, for example, has a Fish Farming Extension Service as part of the national Extension Service of the Ministry of Agriculture, and includes responsibilities for economics and marketing.

Central organization, however, does not appear to work well if additional activities include licensing and policing functions. These are often characteristic of African countries where the prime function of the government department is for the management of wildlife, and can include hunting as well as conservation. In these countries farmers have a distrust of the extension officers. In Benin and Gabon, for example, extension workers also police as game and forest wardens; in Ghana, Kenya, and Nigeria, they enforce fisheries regulations.

Good extension services enable research and development findings to be adapted and applied to industry practice, and to feed back into research and development programmes the needs of the aquaculture industry. A key element in such services is the quality of contact with producers. Extension agents are acting essentially as communication channels between producers and other services, and government programmes in extension should not only include technical training, but training in communications, instruction methodologies, and practical skills for demonstration.

Extension services, except in rare cases, are organized through existing provisions for agriculture and fisheries (see Figure 7.2). In many aquaculturally-developed countries, such as those of southern Europe, there are no extension services organized on a nation-wide basis. Extension is available through research, and research and development organizations which have extension capability but whose primary purposes are in areas other than extension.

The USA and Japan have well organized national extension networks for aquaculture. In the USA extension is organized through the Department of Agriculture, and in Japan it is organized through the regional or prefectural network of fisheries offices and through cooperatives.

Extension in eastern European countries, such as Hungary, has until recently been carried out through cooperative arrangements between research institutes and state farms (see Figure 7.3) rather than through nationally organized extension services.

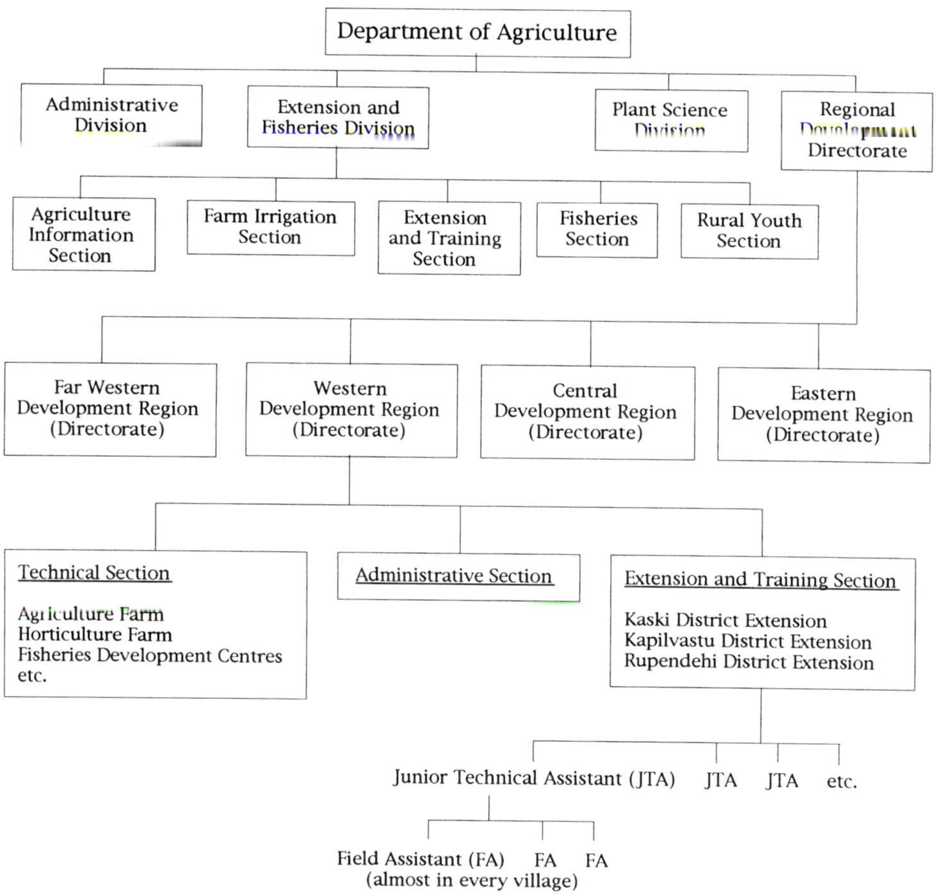

Figure 7.2 Organization of extension for aquaculture through agriculture and fisheries, Nepal (Source: Fisheries Circular FIRI/C747, FAO, Rome).

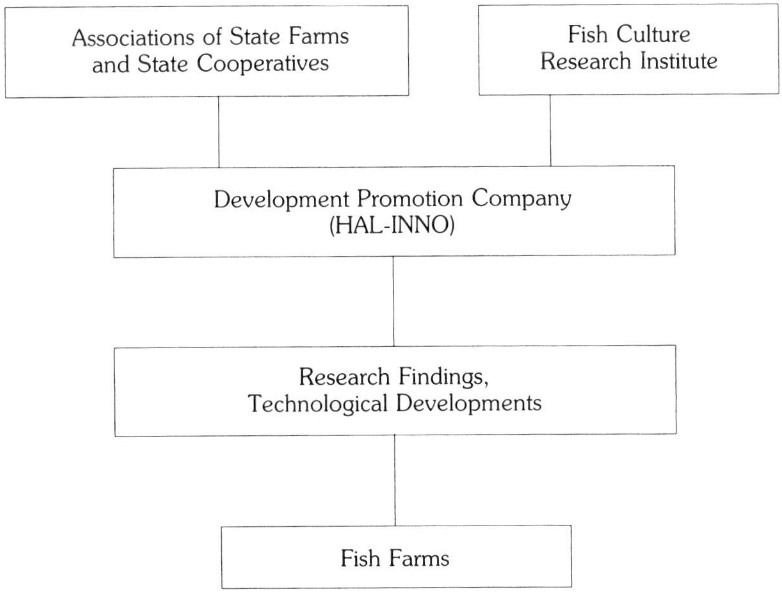

Figure 7.3 Arrangements for provision of extension services to aquaculture in Hungary.

Several Asian countries, on the other hand, such as Malaysia, have established new national extension services for aquaculture, as have several African countries. However, many extension services in Africa are reported to be severely over-manned and under-funded and are subject to excessive bureaucratic control and organized on very hierarchical lines (as in Nigeria).

A close and effective relationship between the producer and extension agent is vital for good extension work. Many services, because of lack of funds and immobility of extension workers, are below standard and of little use for the development of new industries or new methods of production.

Provide any special services

Both the producers and the governments have a particular need for certain special services related to the control of aquatic animal and waterborne diseases for the protection of farm animals and for public health. This includes responsibilities for issuing licences and permits.

The identification and treatment of diseases of aquatic animals (and plants) are responsibilities of the producers, with technical inputs provided by the extension services, the manufacturers of veterinary medicines, and professional scientists involved in research. All may be assisted by special diagnostic laboratories which may be established by the government if the sector is large, or by private enterprise.

Special government services control disease through regulations

Canada provides disease diagnostic services through federal programmes in support of Fish Health Protection Regulations. Federal regulations require a transfer licence for the movement of all broodstock or seedstock of marine fishes and invertebrates, and provincial government regulations require private fish pond permit holders to have a transport permit for the transportation of fish or fish eggs, and an import permit in the case of live cultured fish or eggs from outside the province.

Because of the ready transmission of diseases through contiguous water bodies, or through the transfer and introduction of stocks (sometimes across international borders), there is concern that particularly virulent bacterial diseases, viruses, infectious fungi, and parasites, are not permitted to spread. Furthermore, it is also necessary to safeguard against the spread of vectors of human diseases, particularly those human parasites which

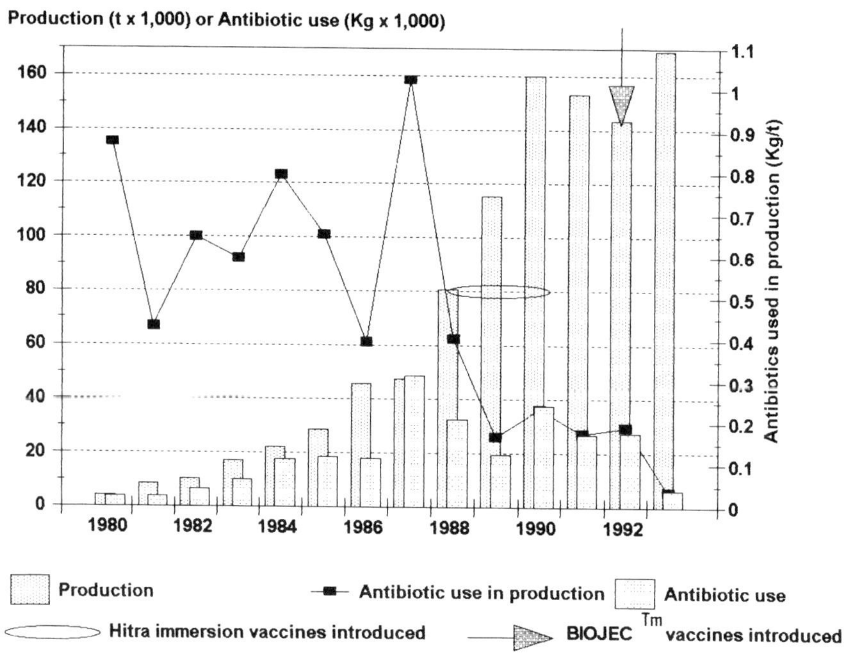

Figure 7.4 Growth of farmed Atlantic salmon in Norway (in thousand t) following the use of vaccines to replace traditional antibiotics to control diseases (Source: Novotny and Nash, 1995).

have an aquatic species as one host within their life-cycles. The economic benefits of vigilance in the control of aquatic diseases, through private or publicly-funded research, are often direct and immediate, as exemplified by the discovery and use of vaccines to replace traditional antibiotic treatments for the principal diseases of Atlantic salmon (see Figure 7.4).

In addition to diagnostic services, special government laboratories may require facilities for quarantining stocks which are being introduced, in addition to the legislative requirements for disease-free certification (as noted in Chapter 6). They may also require facilities for slaughter of dangerously infected stocks, and for disposal.

It is also important that the aquaculture products are safe for human consumption. Dangers to human health may come in the form of certain parasites in finfish and shellfish, contamination of molluscs by bacteria from human sewage, paralytic poisoning by molluscs through the presence of certain planktonic organisms in the water, physiological effects due to the residues of veterinary medicines in the flesh of finfish and crustaceans, the lack of hygiene in processing the products, and poor transportation and storage. These are all standard sanitary control concerns which all food producing sectors have had to deal with in the past, and aquaculture is no exception. Consequently it is important that governments support international codes of practice and include standards of safety for aquaculture products with their other food commodities, and that there are professionals and facilities capable of identifying the risks and enforcing the solutions.

Governments ensure products are safe for human consumption

The Codex Committee on Fish and Fishery Products of Codex Alimentarius is attempting to include a code for the quality and safety of aquaculture products. Components of the code will include product safety and quality, water quality, off-flavours, residues of veterinary drugs, and the public health significance of diseases and parasites. Individual countries have regulations and apply their own quality and safety standards. Italy requires the compulsory depuration in Italy of all live molluscs produced and imported into the country.

Provide insurance services

Insurance for the livestock is another important service for the producers, and for the industry at large as it establishes standards and codes of practice. Insurance for livestock is available from the private sector, but this is mostly affordable by producers of high-value crops, such as salmon, trout, shrimps, and oysters. Producers of less valuable crops invariably require insurance but it is not affordable. Many countries have agricultural insurance schemes for producers of cereal crops, fruit, vegetables, animals, and poultry, but as yet do not included aquaculture crops.

The private sector operates livestock insurance schemes for aquaculture

In the UK insurance for aquaculture livestock has been available since 1974. It is now available through private agents based in Europe, Scandinavia, and North America, and the business is world-wide. Governments operate crop, terrestrial animal, and poultry insurance schemes, but as yet have not included aquatic animals and plants.

7.3.4 Information about the sector

Printed information for the producers is available through extension services, paid for by government, and through producers' organizations, paid for by the producers' membership subscriptions. The information is both biotechnical (new techniques and equipment) and non-biotechnical (prices, marketing information, government legislation, etc.). In countries where the aquaculture sectors are well established there is invariably a secondary industry of information for the trade, including newspapers, business newsletters, magazines, and periodicals. These appear in both national and international languages, and are not necessarily confined to national information.

Publishing information is a large sub-industry of the sector

Norway, with an Atlantic salmon industry of some 5000 primary full and part-time workers, and some 1000–1500 secondary workers, has a peripheral publishing industry which produces seven trade papers and journals in Norwegian, and most farms subscribe to the international trade papers published in English which have information on salmon. In the Philippines, the Council for Agricultural Resources Research and Development produces a series of agricultural extension manuals called *The Philippines Recommends*, which include works on farming certain species of fish and molluscs.

Governments have a specific responsibility to publish information which describes the laws, regulations, and ordinances which apply to the aquaculture sector. This is particularly important in countries where the regulations might be complex, and the producers are required to have permits and licences not only for production, but also for using water and discharging water, leasing coastal land, operating in navigable waters, moving and importing stocks, etc. Information may also be required by the farmers on government programmes in financial assistance (grants and loans), and on their financial liabilities (taxes and duties).

Government has a responsibility to produce information about the sector

Canada has published a bilingual booklet called, *A Guide to Federal Government Programs*, and similar booklets have been produced by the Canadian provinces. Most of the States of the USA have produced guides on the laws, regulations, and permits for aquaculture practices in their respective jurisdictions.

7.4 Common realities of delivering support services

7.4.1 Methods of delivery and use

From a review of the preceding sections, it is clear that there is a wide range of policy instruments available to governments for the regulation, development, and support of any industry. These instruments are normally delivered through existing ministries which deal with principal national industries, general economic development, trade, and education and science.

For aquaculture, almost invariably the ministry of agriculture and/or fisheries has had the prime responsibility for sector organization and management (see section 5.1.3). It has also been normal for its range of responsibilities to include research, training, and extension, but these may have been typically shared with the ministry responsible for education, or education and science. In centrally planned economies, agriculture ministries normally have had control over production.

Research, education, and training functions have typically been discharged through universities, institutes, colleges, schools, special courses, and projects in the field. These have been in turn funded by those ministries responsible for agriculture (or fisheries) and education.

Extension of technology, which is also typically shared between these ministries, has been provided through a wide range of arrangements including:

- special national aquaculture extension services,
- as part of existing agriculture extension services,
- as informal or incidental functions of research organizations,
- through development promotion arrangements such as joint research/ industry projects.

Most ministries of agriculture have frequently provided support for aquaculture through financial assistance schemes. These have been carried out either directly or indirectly through ministry support to state, agriculture, or commercial banks. However, functions related to financial assistance for producers have often been carried out by other ministries which have wider economic and industrial development remits.

Other instruments which have been provided by government departments include market-related services, such as the provision of market information and intelligence, promotion of products, and product development. A range of government departments, including local or regional governments, have been involved in the provision of physical infrastructure for aquaculture.

In providing their support to aquaculture, ministries in some cases have established separate, semi-autonomous institutions. These include state aquaculture development organizations, general development authorities, joint public/private sector companies, parastatal marketing organizations, coordination and consultative committees and, especially in centrally planned economies, state hatcheries, state farms, cooperatives, and marketing companies.

From a general standpoint, the objectives of governments for aquaculture have been pursued through the use of instruments which are already in place for other industries or sectors of the economy. This generalization applies, to financial measures; most grant, loan, and tax exemption schemes for aquaculture have benefited from having first been designed for agriculture and fisheries.

The administration of aquaculture has normally been carried out by a subsidiary section of a fisheries directorate in a ministry of agriculture. Research institutes or organizations with specialized aquaculture remits are rare, and extension services for aquaculture have usually been provided through the existing networks or arrangements which have been established for agriculture or fisheries.

Separate and comprehensive provisions and services for aquaculture are rare. One exception is the National Aquaculture Enterprise of Cuba, which is an instrument designed to perform the complete range of aquaculture development and support functions.

In general, the problems associated with the delivery and use of policy instruments for aquaculture have been essentially the typical ones of bureaucratic control and management. The need for production and marketing functions to be removed from direct control by government departments has been apparent in both free-market and centrally planned economies. Finally, the effectiveness of policy instruments for aquaculture has been very dependent on the context within which they have been applied. Accurate assessments of the constraints on development and the application of supporting measures to cover all the needs of the industry are essential if an instrument is to have a positive effect.

7.4.2 Other associated problems

There are a number of common problems which have been experienced in the use of the foregoing policy instruments provided by governments. These are discussed briefly in the following paragraphs.

In the case of financial assistance and related schemes, major difficulties have been created for the developer through slow administrative procedures and excessive bureaucracy from the government departments responsible for their administration. The multiplicity of departments involved in aquaculture support has often resulted in a lack of coordination, duplication of effort, and inter-departmental competition. This has been inefficient and wasteful of resources.

In general, government departments have been unsuited to the production and marketing areas where rapid management decisions are necessary. In recent years there has been an increasingly wider acknowledgement of the difficulties of direct management of these aspects by government departments. This is true of both free-market and centrally planned economies. In the latter, the problems of bureaucratic control of production and marketing functions have to some extent been reduced by the use of a separate 'state enterprise' status which is given to organizations involved in these areas.

Common problems in the provision of government support services of research, education, training and extension have been the lack of applied and practical content, under-funding, poor facilities and equipment, and a generally poor match between the provisions of the services and the needs of the industries. The fact that there were fewer economic or technical barriers to public investment in these services resulted in over-manning and over-provision, and a subsequent loss of creditability with the industry. The provision of infrastructure for aquaculture has often been difficult and costly because of the remoteness of locations in which aquaculture development takes place. Competition for public resources for investment in infrastructure has therefore been very strong.

Institutions set up by government, such as parastatal marketing organizations or development authorities, in many cases have been rendered inefficient through excessive government interference, and by poorly defined or over-ambitious objectives set for the organization. Being government appointed agencies, the boards of these institutions have often been dominated by people with insufficient experience of industry to direct and manage effectively an organization which has to be geared to the needs of industry.

Governments, for the most part, have also failed to produce and publish information for the sector concerning legislation, regulations for permits, financial schemes, and technical advances.

The benefits and effectiveness of government policy instruments have been primarily the building of the capacity of the sector and attainment of objectives which would otherwise have not been reached or at least delayed. The creation of national capabilities in all areas of aquaculture has been dependent on investment in the schemes and services, organizations, facilities, and infrastructure outlined. These were long-term investments with a relatively indeterminate value, and normally without any short-term commercial value. They were not likely, therefore, to be undertaken by private investors without incentives and support, and government investment consequently was essential.

The benefits of each particular instrument are difficult to discern or to generalize, as each has been applied in a wide variety of circumstances and with varying degrees of skill. Financial assistance and associated schemes have had the positive effect of stimulating investment in new forms of production, or in low-investment regions or populations. They have been highly effective in 'pump-priming' developments, which private and commercial investors are normally unable or unwilling to undertake. Variations in the rates of assistance has helped to reduce the natural imbalances or inequalities involved in the economy by directing economic opportunity to disadvantaged areas or groups. When they have been applied as compensation or insurance schemes, they have helped to reduce the effect of accident or disaster.

In centrally planned economies, subsidization of inputs is a part of the general economic system which involves price control according to politically determined (as opposed to market-determined) criteria. However, subsidization of inputs has also been applied to help increase food supplies in low-income economies, or in emergency situations where there was neither availability of finance nor experience of its use or management.

The benefits of government services and supports have been in the development of human resources within the country, and the increase in national capability which this created. Specifically, investment in research has created a greater capability to innovate and to develop new forms of production and adapt existing ones. Investment in education and training has resulted in the provision of skilled manpower for the industry. Institutional support has been designed to make government effort more effective and relevant to industry's needs.

The question of the effectiveness of particular services or groups of services is essentially a matter of the context within which they have been applied. A clear illustration of this has been Cuba's use of a parastatal organization, the Cuban National Aquaculture Enterprise, to perform all the functions, from research to production and marketing, of all forms of aquaculture. In the particular case of Cuba, with little experience of aquaculture and, consequently, with a comprehensive range of development needs, an independent and specialized organization with comprehensive functions has been highly appropriate and cost-effective within the framework of a centrally planned economy. However, in larger countries with established aquaculture industries and limited development needs, the more effective provision, from a cost and operational point of view, has been the use of support networks already in place for fisheries or agriculture, and these have been adapted to address the more specific needs of aquaculture in those countries.

Similarly, certain policy instruments have had little effect on development where they were not part of a general provision of support. This accurately reflects the true nature and scope of the constraints on development. Thus, for example, a tax exemption or grant and loan scheme was not entirely effective as a means of stimulating development if the government was unable to provide land or access to suitable development areas. A more specific example has been the over-manning of extension services in some African countries. Thus there was little point in providing support services if the industry could not develop because of the lack of trained personnel, finance, or essential physical input.

In summary, where they have been most effective, government services and policy instruments were selected and designed for particular circumstances. They were based on an accurate and continuing assessment of the true constraints affecting the target industry. Moreover, the effectiveness of any one instrument was also dependent on whether others were in place, thus catering for the needs of the sector as a whole.

Chapter 8
The Use of Policy Instruments III – Development Projects

Synopsis

This chapter describes the third type of policy instruments available to governments to manage and build the national sector, namely the use of development projects. It begins with the role of projects in sector development in general terms. It describes the many types of projects, and their limitations, which may be considered, most of which are supported by examples or larger illustrations. This is followed by a description of various methods of financing projects, and concludes with variations for their financial arrangements.

8.1 Projects and their role in sector development

8.1.1 Introduction

Development projects have described a wide range of activities in terms of complexity, technical requirements, and scale. All of them, however, have shared a common denominator. Those investing, whether individually, within a cooperative or a company, or from within a government agency, have made the investment in the anticipation of receiving in the future a financial return on their investment, or other benefits to the economy which may or may not be susceptible to quantification.

Many aquaculture development projects have been concerned with the production of aquatic animals and plants, or their marketing. Such investments are of a business nature and, in most cases, those making the investment are looking for a financial return. In addition, these projects are expected to generate other benefits to the economy, such as the creation of new jobs, increased net foreign exchange earnings or, perhaps, increased supplies of products to the domestic markets.

Marketing projects, which are often closely associated with production, have included investment in a processing plant for frozen, smoked, and canned mussels in Mexico, a processing and packaging plant specifically designed for eel products for export from Taiwan, or marketing trials in the UK of new farm products from Thailand.

Production investment projects have included, for example, artisanal farming of tilapia in the Philippines, clam farming in France, semi-intensive marine shrimp farming by joint-venture enterprise in China, or an intensive offshore floating cage farm for salmon in Norway. While most 'business' type projects like these are 'near market', other production projects are a little further away, for example, the investment in a hatchery in the Central African Republic to propagate catfish seed, or in a shrimp feed plant in Ecuador; and yet others are even further away, such as investment in a factory fabricating fibreglass tanks and seacages in Norway, or a company to publish a trade paper in the USA.

As aquaculture is a relatively new sector, most projects have been concerned with the provision of physical infrastructure to enable producers to function more effectively. Examples of such projects are, for example, the expansion of aquaculture in central China, with extensive land drainage and construction of feeder canals to provide new areas for farming, construction of roads, provision of utilities, building or improvement of markets, and so on. Other projects have been designed to assist a target group of beneficiaries, such as the provision of a line of credit for shrimp farmers in Bangladesh. In both these cases, where investment can be related to an estimated future level of production or higher market values, it is possible to estimate the economic benefits which will be derived from that investment. Clearly, however, as the return consists of benefits to the economy rather to the individual investor, the investment has been made by government.

Finally, there have been many projects designed to advance and strengthen the sector in general, rather than a specific group of beneficiaries. Examples of such projects have been, for example, research and development of salmon farming in Chile, graduate and post-graduate education for specialists in the Philippines, the provision of a legislative framework for aquaculture in Canada, or training for civil servants in the planning and management of the sector in Nepal. Such projects are so broad that it is unrealistic to attempt to quantify the benefits which will be derived from them, other than through a 'least-cost' analysis in their appraisal. Like projects concerned with physical infrastructure, these general projects have been made by government investments.

All of these activities have required financial outlays and imply risks for the investors, whether public or private resources are at stake. For most investments a prerequisite of funding, whether it be from an individual, a parent company, a bank, a government department budget, or from a partnership between them, has been the preparation of a well-designed project document to ensure that the project is fully understood and analysed to provide the financial or economic return required.

8.1.2 Development projects and their geographic boundaries

In general, the size of any development project in geographic terms frequently influences the number of 'partners'. Projects, independent of their sub-sectoral objectives, often extend beyond a single site to different sites within the same area, the whole country, or even several countries. In brief, there are four types of projects described by their geography. These are:

Single site projects, which may range from the artisanal farm unit, with production of a few tonnes, to local but complete aquaculture development of a valley, lake, lagoon, or bay, producing collectively thousands of tonnes and employing many hundreds of people within a cooperative farm system and/or companies. Such projects are generally oriented toward production of one species or species group, and often include some form of communal organization for land and water use, purchasing supplies, and marketing products. Typical single site projects have been, for example, private investment in a salmon farm in Scotland, or an integrated poultry–fish farm in Thailand.

Local projects, which may involve many production units on different sites within the same zone. These may have one or more organization and management responsibilities, often with a marked geographical discontinuity between them; such as production practices requiring different water sources for hatchery production and grow-out phases, or rehabilitation and/or production improvement projects through better feeds and training in feed utilization. Examples of local projects have been rice-cum-fish farming for a group of traditional rice farmers in Vietnam, or the introduction of fish farming to the artisanal fishermen of the coastal lagoons of Turkey.

National projects, which may extend throughout several provinces where production is concentrated, and possibly throughout the country. Such projects include many different activities and have components of production, demonstration, extension, training, research, processing, marketing, and financing. These projects are generally targeted at specific beneficiaries (such as the rural poor) to take advantage of new aquaculture technology and its opportunities. Typical national projects have been, for example, development of fish production in the Terai of Nepal, and in the Côte d'Ivoire.

Regional projects, which may extend throughout several countries or all countries in a geographic region. These are mostly technical assistance projects with objectives to improve the exchange of information and dissemination of technology through components of services, training, management, and financial support in the acquisition of foreign equipment. Examples of large regional projects have been for the countries of the Mediterranean through the Mediterranean Regional Aquaculture Development Project, and the countries of South and Central America through the Aquaculture Development Project for Latin America.

As the projects increase their geographic size, there is a corresponding increase in the number of federal, state, and local government administrations which have to be involved to some degree. In the case of regional projects this will be a group of different governments, and probably several administrations within them.

8.1.3 Development projects in relation to sector capacity

In general terms, projects vary in accordance with the state of the respective national aquaculture sector, and its capacity to handle development. Unfortunately many countries have had no aquaculture sectors at all or, at best, the national activity has been made up of a number of independent research and development activities. Consequently many projects, particularly those of the private sector, have been investing in a venture where no aquaculture activity has existed before. This is one of the major differences between aquaculture projects and those of fisheries and agriculture, which may also be new but are a part of a large and active national sector.

As a result, it is still common for aquaculture planners, whether in the public or private sectors, to find themselves in the position of pioneers. In order to carry out a development project they have had to explain first of all the basic principles of aquaculture to local authorities and banks, to counter negative opinion about other badly managed or unsuccessful projects, and then initially to demonstrate their concepts on a small scale. Fortunately, in the last decade the benefits of good aquaculture have become more widely known. Aquaculture is now more widely practised, and therefore there have been more opportunities provided for projects to improve production and rehabilitation.

Consequently aquaculture projects fall into three categories:

- Projects intended to create the sector,
- Projects to improve production of the existing sector, and
- Projects for rehabilitation of the sector.

Projects intended to create the sector. These are concerned with developing the capacity of the sector in areas where it has not existed before. In addition to constructing new farms, such projects may concern introduction of new species or new technologies to exploit opportunities. For example, in the last 15 years many countries, such as Ecuador, Panama, the Philippines, and Thailand, have introduced commercial hatcheries for production of varieties of marine fish and shrimps, which have subsequently exploited development opportunities for export earnings.

Projects designed to improve production of an existing sector. These are particularly important for aquaculture development at the present time, especially countries in Asia, such as China, India, and Indonesia, where there are long-established traditions but techniques and yields are far below what the state of the art can permit. Intensification of production has been achieved through projects with components of management, technical training, extension services, better feeds, and improved engineering and machinery.

Projects for rehabilitation. These are important to a non-productive or damaged sector. Such projects have long been a tradition of freshwater fish culture in African countries, such as Egypt and Nigeria, where the potential economic returns have been so limited that the inputs of seed production and extension services have been entirely left to the public sector. The lack of financial resources has invariably led to mismanagement, insufficient support to the farmers, and finally complete loss of interest. The need for rehabilitation projects remains, both for individual farms and large areas of farms. Fortunately financial interest by both the public and private sectors is returning due to prospects with new species and technologies, and the growing competition for sales of exportable commodities (such as marine fish) on international markets. Furthermore, there is growing ecological pressure on aquaculture to avoid impacts (such as the spreading of diseases, and increasing nutrient loads) which justifies emergency projects to rehabilitate damaged aquaculture sectors of consistent economic importance.

8.1.4 Development projects and the technical parameters of aquaculture

Projects, whether small or large, vary according to a number of technical parameters. The major variations are due to the number of species involved, the types of production technologies used, the target groups in terms of intended beneficiaries, and the government administrations which may have technical responsibilities for different components. There are four types of projects characteristic of differing technical parameters:

- Projects involving only one species,
- Projects based on a single technology,
- Projects which address a single target group, and
- Projects which involve one or more administrative bodies.

Projects involving only one species. These are usually small individual enterprises. When they have been large, involving many producers together, they have been based either on production of a low-value species for increasing national consumption of cheap protein (for example, mass production of tilapia in Côte d'Ivoire or Haiti), or production to increase national income through exports (for example, marine shrimp culture in Indonesia). Most projects have been built around options for several species, the culture of which has either been carried out synchronously in the same volumes, for example, polyculture of Chinese and Indian carps in earthen ponds in Nepal, or linked by a common technique or environment, for example, culture of shellfish (oysters, mussels, and clams) suspended from rafts or on beds in coastal areas of Malaysia.

Projects based on a single technology. These have been characteristic of bilateral organizations and the private sector, both of which are interested in demonstrating and using a particular technology. Such projects have involved one or a group of related species, and one particular phase of the production process or production requirements, for example, the improvement of smolt supplies and feed technology for salmon production in Chile. Projects based on several technologies have been mostly for comparison and selection of the most appropriate technology for the area concerned, or for demonstrating a range of options for technological transfer between partners. They have been more characteristic of regional cooperation projects where one key objective is to facilitate dissemination of technologies from one country to another, for example, the introduction of technologies for sea bream, sea bass, mullet, oysters, and mussels to a group of Mediterranean countries by the regional aquaculture development project.

Projects which address a single target group. These projects have often been addressed to the rural poor or, in some cases, one particular social category, such as women fishfarmers in Malawi and Zimbabwe, or displaced fishermen in Egypt. Such projects are becoming less frequent. Regional development banks in particular have redirected the emphasis of projects from small-scale subsistence production to commercial projects which generate development of small-scale farms in the surrounding area, for example, the concept of the nucleus estate, as practised by private companies and small-scale shrimp farms in several countries in Asia, such as the Philippines, Indonesia, and Thailand.

Projects which involve the participation or clearance of one or more administrative bodies. Invariably these are projects under one responsible ministry (such as agriculture or fisheries). However, as aquaculture is not recognized anywhere as a sector deserving of its own ministry, for technical reasons most projects have involved the participation of several such bodies (such as ministries of agriculture, fisheries, forestry, departments of animal production, water resources, and environment, etc.). For example, projects in many Latin American countries, such as Brazil, with components of research, training, and production, have had to deal with a number of administrative partners and governmental agencies.

In conclusion, development projects in aquaculture have differed according to the category of project required by the current capacity of the sector, and the technical parameters concerned. However, all projects have to be financed by someone, and effort committed to their operation and management. These are considered in the following section.

8.2 Financing aquaculture development projects

8.2.1 Introduction

This section describes the major types of investment which have been made to finance development projects in aquaculture, and some of the possible variants. As the project approach to development depends fundamentally on the mechanisms used by the promoters (both public and private) to finance their types of projects, the section begins with a general description of the options available for investment, which in turn categorizes the different types of projects into a structured framework.

In any development project, investment is necessary for three categories of acquisitions. These are:

- Depreciable capital facilities and equipment, which serve a project without being destroyed or transformed themselves in the production process, such as hatcheries, ponds, tanks, and pumps,
- Material goods which are consumed or transformed in production processes, such as fuel (to pump water) and feed for livestock,
- Non-material goods, such as technological knowledge, and all forms of services.

All are potential basic expenditures to be borne by any project, but they are treated differently by financial institutions as they are very different in terms of associated financial risks. This dictates the variety of approaches to project investment, and the classification which has been developed.

Investment in development projects in the aquaculture sector comes from two principal directions, namely external sources and domestic sources (see Figure 8.1).

PROJECT INVESTMENT CAPITAL

External Sources	Domestic Sources
Concessionary loans Capital grants Equity participation Other aid schemes Direct foreign investment	Owner capital Company capital Share capital Bank loans Loan assistance schemes Government grants Government subsidies

Figure 8.1 Financing development projects.

External investment for projects is available in the form of:

- 'hard' and 'soft' (concessionary) loans provided by multilateral development agencies (such as the World Bank, EU), regional and bilateral development banks, and multilateral and bilateral aid organizations;
- capital 'grants' provided by multilateral and bilateral donors;
- equity participation by both banks and aid organizations with loans on a joint-venture basis; and
- direct foreign investment.

Both hard and soft loans are utilized normally through lines of credit made available from the external financing institution to a financing institution in the recipient country. When the loan or grant is used to provide credit to individual farmers, the national financing institution usually provides a line of credit to an agricultural development bank which makes the individual loans. A characteristic of both hard and soft loans is that the external financing institution is usually responsible for providing the foreign exchange requirements of a capital investment, while the recipient country provides the domestic financing requirements. This pattern does not always apply, however; exceptions usually occur when domestic currency expenditures form a large proportion of a relatively large investment.

Capital grants are given by multilateral donors or bilateral donors. As the name implies they do not require repayment by the recipient country but their individual size is often, although not always, relatively small. Technical assistance is often provided from the source of capital. Its purpose is usually to enable the recipient country to make the best possible use of the capital investment. It usually forms part of the total loan or, when capital is provided in the form of grant aid, it is usually provided on similar terms. In addition, technical assistance is often provided as a direct entity, not related to a capital investment.

Many banks and certain aid organizations offer loans through equity participation, or joint-ventures. These are usually short term at concessionary rates, and repayable in full once the business is viable, or converted to a grant if the business proves not to be viable.

Foreign investment comes in the form of independent holdings, direct purchases of national companies, and joint-ventures.

Domestic investment in projects comes from the private sector, either through privately or publicly-owned companies or through individuals, and from the public sector. These are usually supported by bank financing. In the case of domestic investment a number of quasi-commercial loan assistance schemes are available from development banks within developed and developing countries, as well as grants and subsidies.

8.2.2 Projects financed from external sources

Hard loans

In the two decades of aquaculture development, which is embryonic compared with modern agriculture, the practice of all the external assistance organizations, including the banks, has been to provide generous support to get the new sector established. Therefore few if any hard loans have been made directly to the sector for development projects. The only instances of hard loans have been where aquaculture projects were included as components of large national loans which had relatively short-term high-interest rate conditions. This may have been because recipient countries had a high GNP and were semi-industrialized, and were rated within a 'hard' loan category assessed by the World Bank, or because the majority of components in the loan package were commercial in nature.

Hard loans are provided at the national level through financial institutions

As part of a US$135 million 'country programme loan' by the World Bank to Chile (i.e. a comprehensive loan for one or more national sectors), a small component (below US$500 000) was for salmon ranching. Chile is ranked by the Bank as a country with a 'middle-income economy'. Commitments by the Bank of US$754 million in 1986 had an average interest rate of 8.6%; average maturity 19 years, and average grace period of four years.
Sources: ADCP/REP/89/39, FAO, Rome, and the World Bank, 1988.

Soft loans

The greatest contribution to aquaculture development has been made through soft or concessionary loans from the multilateral development banks, the bilateral development banks, multilateral donors, and bilateral aid agencies. These loans are for projects which invariably have components of capital aid, technical assistance, and a line of credit.

As the World Bank prefers to make large programme loans to countries for development, the support to aquaculture is invariably only one component among a number. For example, a typical aquaculture component is from US$5–30 million, in a loan of US$200 million or more, and includes elements of hatchery and pond construction, training, credit for farmers, and management costs. The loan may be part of general rural development, or part of a substantial construction project for water utilization, such as irrigation reservoirs and canals, or hydro-electric dams. Often the loan is combined with a domestic investment programme by a recipient national bank.

The regional banks, on the other hand, such as the African, Asian, and Inter-American Development Banks, have made individual and substantial loans to the aquaculture sector through the national agricultural development banks for projects to finance construction, and provide a line of credit, technical training, and project management.

Most loans to the sector are concessionary

The National Bank for Agriculture and Rural Development in India (NABARD) provides credit refinancing facilities to all state cooperative banks, land development banks, and commercial banks including regional rural banks. Depending on the type of proposed project and area of operation, NABARD refinancing ranges between 75–95% of the total bank loan. In the fisheries sector it caters to the long and medium-term credit requirements. Bank loans can be given to individuals, groups of progressive entrepreneurs, cooperative societies, fishermen's federations, state fisheries development corporations, and also private companies. Subject to suitability of schemes, NABARD's refinance is made available to any category of above banks at the rate of 6.5–8%. Beneficiaries pay a rate of 10–12.5% interest. While small farmers pay interest at 10%, others get a loan at 12.5%.
Source: Pathak & Mohanakrishnan, 1987.

Some concessionary loans may be directly at the producer level; for example, a Small Investment Credit Programme in Indonesia provides loans for intensification and extensification of brackishwater and freshwater ponds for aquaculture. Eligible applicants have maximum credit of Rp15

Several United Nations organizations offer concessionary loans to governments

The International Fund for Agriculture Development (IFAD), jointly with the Asian Development Bank (ADB), co-financed the second phase of a US$20 million Aquaculture Development Project in Nepal, executed by the Department of Agriculture and the Agricultural Development Bank of Nepal (ADBN). The IFAD component was for approximately US$5 million, with terms of 50 years including a grace period of 10 years, and service charge of 1% per annum. ADB co-financed with US$10.5 million, with terms of 40 years, 10 years' grace, and 1% service charge per annum. UNDP made a grant of about US$1 million to the project. The Government re-lent US$6.75 million to the ADBN for the credit component under certain terms and conditions (15 years, and five years' grace). ADBN onlent to small-scale farmers at usual interest rates, which are about 12% for fisheries activities and 15% for marketing activities.

---

million at an annual interest rate of 12%. The loan period is 10 years, including four years of grace. There is a 3% penalty for loans in arrears. Loans from commercial banks are 2–3% per month, and are short-term with no grace period.

Certain United Nations (UN) organizations offer long-term soft loans for development projects, in association with other assistance. Some of these have been active in the aquaculture sector. For example, the International Fund for Agriculture Development is financing aquaculture construction projects, particularly pond complexes, through loans which have long-term credit and low-interest.

Capital grants

In the developing countries capital 'grants' for projects have made a significant contribution to establishing, and in several cases creating, their national aquaculture sectors. Such grants have been provided by the multilateral and bilateral donors and have been for (a) capital aid, and (b) technical assistance. Although both are invariably given in combination much depends on the limitations of the donor.

The principal multilateral donor has been the UN system, but increasing contributions to the sector have been made by the EU. Most of the organizations of the UN system are restricted in the form of their assistance, either capital aid or technical assistance, and some to only project execution.

The major bilateral donors to aquaculture (in alphabetic order) have been France, Italy, Japan, and the USA, followed by Australia, Canada, Denmark, Germany, the Netherlands, Norway, Sweden, and the UK. The bilateral donors are not necessarily restricted for their project support.

Capital aid projects for construction of all types of aquaculture facilities have so far played a vitally important role in development. Principal facilities qualifying for capital aid are national research centres, government demonstration farms, production hatcheries, and small-scale pond complexes for new farmers. Depending on the agency or financing organization, capital aid is usually given in association with technical assistance and/or a component for credit.

Hatcheries and pilot/demonstration farms have probably been the most useful contribution to development through capital aid projects and these are typically small investments (below US$1.5 million) financed by the bilateral organizations. However, there are some exceptionally large grants made at times, often because of preferential political reasons.

UN organizations which cannot directly provide capital aid, such as UNDP, have linked their technical assistance with other multilateral and bilateral donors who can, or with national investment.

Bilateral agencies make many capital grants for facilities and operational assistance

The US Agency for International Development (US AID) provided capital assistance of some US$30 million to Egypt to construct a National Aquaculture Centre (complete with laboratories, hatchery, and research ponds), residences for staff, dormitories for trainees, a model homestead village, and the construction of over 600 ha of ponds together with canal systems (first phase). In addition US AID provided technical assistance with over 15 man-years of expatriate expert consultation, and over 60 man-years of education and technical training in the USA. It also provided extensive equipment for site operations, and technical equipment for all laboratories

Technical assistance projects

Technical assistance projects, with or without capital grants, through the multilateral and bilateral donors have been the most significant factor in the

International technical assistance provides support to the sector at all levels

- At the regional level the government of Italy provided US$4.5 million for aquaculture development in the Spanish and Portugese-speaking countries of Latin America through a four-year project executed by FAO. The project provided technical training courses in planning, fish nutrition, economics, lagoon management, etc., established research networks, and provided information to 19 countries.
- At the national level the Overseas Development Administration (UK) funded a project with UK£1.4 million for technical assistance to support the Mangalore Fisheries College in India with the help of two British universities. Equipment was provided for research in aquaculture and fisheries activities, together with a component for training.
- At the individual level the International Development Research Centre of Canada (IDRC) has provided innumerable overseas fellowships for education in aquaculture disciplines relevant to IDRC's programme interests. These were therefore for selected candidates, eligible for post-graduate qualifications. The International Foundation for Science of Sweden has provided research support in-country for qualified individuals working on aquaculture topics. Current research grants are for between US$10 000–15 000, and usually for facilities, equipment, and materials only, rather than a stipend.

development of the aquaculture sector in most countries. Usually technical assistance has been tied to national capital investment, but this is not always the case.

Technical assistance through both loan programmes and individual projects has brought in international experts to advise, operated facilities, purchased equipment, educated manpower, trained technicians, and provided technical information. In many cases long-term technical assistance has built the entire aquaculture capacity of the country.

Individual technical assistance projects are mostly national, and may range from a number of small research fellowships and study tours, to long-term fully staffed national and even regional projects.

A number of agencies, such as the International Research Development Centre from Canada and the International Foundation for Science from Sweden, support projects only in education, training, and research.

Other investment avenues

There are other opportunities for financing assistance projects. The World Food Programme, for example, has financed projects indirectly through food aid in exchange for work in the construction of infrastructure, such as canals, ponds, and hatcheries.

Other assistance is provided by International Food Aid

The World Food Programme (WFP) supported a US$63 million project in China with US$38 million of which US$34 million was in food aid. The project involved the construction of 5466 production ponds and 1366 ha of fingerling ponds in nine different municipalities in an attempt to triple freshwater fish production. A total of 65.5 million work-days were estimated for the construction over a three-year period. WFP assistance provided budgetary support to the government by paying part of the wages of skilled workers and contributing grain in lieu of subsidies provided by the government. It also obviated the need to buy additional grain to meet requirements. In addition to the construction of the ponds and extra production of fish, the project also developed low-lying flooded lands which were unutilized, and created temporary and long-term employment opportunities thus raising the local economic status of the depressed area.

A number of multilateral agencies of the UN system, such as FAO and the UN Educational, Scientific, and Cultural Organization (UNESCO), have executed projects which linked their expertise to domestic investment. Both have used trust funds provided by the host country to construct national aquaculture research and development centres, hatcheries, and farms.

The use of special funds for development assistance

- The government of Saudi Arabia provided US$2.9 million and SR21.25 million (US$5.6 million) funds in trust for the construction of a fish farming centre at Jeddah, and technical assistance. The trust fund agreement with FAO was for the engineering and technical advice for the design of the centre, for resident international experts to direct and advise relevant research, and for study tours.
- Similarly the government of Oman, through an agreement with UNESCO, provided funds in trust for the construction of a fisheries research station (including some aquaculture), and resident international experts to establish an institutional structure for fisheries and direct relevant research activities.

Another investment avenue has been provided by FAO through its own Technical Cooperation Programme (TCP) for projects to assist developing countries to plan and develop larger aquaculture projects in anticipation of subsequent funding by other donors. For example, in anticipation of further technical assistance through UNDP to Zimbabwe, FAO provided assistance of US$187 000 from its TCP funds to set up an extension service to promote aquaculture in rural communal areas, and to provide training of extension personnel. The project also included rehabilitation of the Henderson Fish Farm, to be the focal point of the work.

It has also been possible to obtain support for aquaculture development as part of assistance to a broader but related field. For example, UNESCO, in its support for marine science research and education has included components of fisheries and aquaculture in its purview. With national funds, it has carried out projects in several countries to construct facilities in support of national aquaculture development.

Direct foreign investment

Foreign direct investment in projects has rapidly become the most practical approach to development. A number of large foreign corporations have bought ongoing and successful farms outright, where they have been permitted to do so; for example, a Japanese consortium which produces and markets seafoods, recently paid US$22.1 million for a salmon farm, complete with all stock and on-going operations, in Chile. However, the most common form of investment has been through joint-ventures, with foreign enterprises typically providing up to 49% of the equity.

There have been two typical formulae for joint-venture projects, both based on relatively simple premises:

(a) foreign capital is usually guided by the respective national strength in one form of technology, and therefore seeks joint-ventures with companies in countries where there are conducive resources; for example, because of national policy restrictions on the size and number of salmon farms in Norway, Norwegian companies have been investing abroad through joint-ventures in countries well-suited to salmon farming. Over NKr 500 million was transferred through currency permits handled by the Norges Bank (pre-1987) for investments in Canada (27%), Iceland (24%), UK (16%), USA (8%), and other countries (25%).

(b) foreign capital is guided by respective national strength in markets for the products, and therefore seeks joint-ventures in countries where these products can be produced; for example, enterprises in the USA and Japan, where there are the world's largest markets for marine shrimps, have joint-ventured in the Philippines, Indonesia, and Thailand, where marine shrimp can be produced in ideal conditions.

However, joint-venture projects have not always been based on capital investment. Although both investors contribute to the equity of the partnership, the equity has often been based on contributions of technical knowledge, expert manpower, and training. There have also been arrangements for distribution of the profits through sales of product outside the country, which is usually the method for foreign investors to transfer the profits home. Similarly, joint-venture projects have not always been the prerogative of the private sector. A number of quasi-governmental organizations have been active in aquaculture and enter into joint-ventures.

Foreign joint-venture development between the private and public sectors

The French government (FoG) and the Territory of French Guyana (TFG) joint-ventured in a project to form a public/private company to construct and operate a shrimp hatchery and provide services to farmers. The basic cost of the capitalization project was about US$5 million. The company was incorporated and capitalized by FoG, through the Institut Français de Recherche pour l'Exploitation de la Mer (IFREMER), and by TFG, through its National Agronomic Research Institute, for 20% of the project requirements. The company was then eligible for subsidies of a further 50% from the region, French community funds, and the EU, and loans (30%) from the regional development bank to private farmers through their federation. In addition to engineering design, construction, and operation of the hatchery, the company carried out extension and training for the farmers, using the expertise of IFREMER, and research, using the Agronomic Institute. The company also assisted the large-scale commercial farmers on a private basis with services, particularly further financing and loan processing with the EU, and services for small-scale farmers were provided for financing from public funds.

8.2.2 Projects financed by private investment

The majority of projects in aquaculture world-wide are financed by individuals or private companies using their own financial resources, or resources borrowed from commercial banks, or by incorporated public companies using resources raised through the sale of stocks and shares.

The greatest contributors to domestic investment have been the individual entrepreneurs. Most farms are smallholdings, owned and operated by one man, and possibly his extended family, having invested money, time, and effort in construction. Typical pond construction costs are US$500–3000/ha, depending on site location. The majority of credit available through commercial and agricultural development banks has gone towards pond construction, although in developing countries many rural farmers prefer non-institutional credit from traditional village moneylenders.

The next largest group of contributors has been the company investors. Many farms, both large and small, have been owned privately by an individual or groups of individuals, having increased their assets over many years through personal investment, assisted by commercial loans and grants. The group making the smallest contribution have been the corporate investors, although a number of individual investments are extremely large. Many large farms for marine shrimp and salmon, for example, are owned by incorporated companies. Stocks in some of these companies are publicly traded on stock exchanges in the USA, Canada, and Norway. However, these are all large corporations and their investments in aquaculture are minor activities of diversification, usually through purchase of going concerns. For example, Norsk Hydro, the principal Norwegian energy company, owns and operates salmon farms; and British Petroleum (until recently) owned farms as well as being a major feed manufacturer with plants in many countries.

Private investment has been assisted by a number of support services which take the form of government programmes and projects. These have been financed by the national treasury, or through hard or soft loans from international development banks and commercial banks. The repayment of these loans by governments is therefore a further important domestic investment in the national sector, established by government policy. Typical support services include:

- quasi-commercial credit assistance and insurance schemes,
- grants and subsidies, and
- indirect support services through administration, extension, research, education, and training.

These have all been reviewed in detail in Chapter 7. However, variations in these financial arrangements will be described in the final section in general terms.

8.3 Variations in the financial arrangements of development projects

Although the framework for financing development projects outlined in the previous section is relatively simple, individual projects have been complicated to some degree by a number of variations, such as restrictions and the number of project objectives.

8.3.1 Restricted projects

It is important to know what portion of the value of the project and its benefits accrue to the original project donor or lender, taking into account the internal profitability of the project itself.

For any form of *capital grant* the donor does not expect any direct return but finances, and at times manages the project without imposing restrictions on the rights of the grantee. Capital grants have been typically provided and executed by international organizations, such as UN agencies and governments of developed countries with bilateral assistance programmes. Invariably consultants have been recruited on the international market by the donor with approval of the recipient. Equipment and all consumable materials have been purchased on international markets, again with selection and approval of the recipient. In addition, there has not usually been any restriction on the publication of the results of the project, and no limitation on the use (including sale) of technology acquired through the project.

For projects financed through private investment (particularly by foreign investment), with or without credit, all or nearly all the profits of the project have returned to the original source of investment, except the host country or region has also benefited from the wealth of labour generation. In the private sector this type of project describes joint-venture investment with 'buy-back' agreements, in which the foreign partner has provided finances and technology (with conditions of confidentiality), and is directly or indirectly, through a subsidiary company, the sole buyer of the products. The host, or local partner, has provided the land and the local labour force. Such projects have been common in shrimp culture, particularly in agreements between international holdings diversifying into other production opportunities and local authorities committed to increasing employment opportunities.

In both these cases there may have been restrictive conditions imposed by the grantor, the lender, or one of the partners. These may have been to recover a predetermined part of the project profits and benefits, either for their own direct interest or for whom they have responsibility. For example, in the public sector the lender or grantor may have restricted international involvement. Regional development banks, the equity of which has been built by a limited number of governments, may have restricted the use of consultants and/or equipment only to its member countries; similarly regional development funds, such as the European Development Fund, may have restricted in the same way the resources of technology and materials. The bilateral organizations, such as those of Germany, France, the UK, and the USA, and several other cooperation agencies, have usually restricted recruitment of consultants and the supply of equipment to companies from their own country, while Italy and Japan have required that equipment was manufactured in their own country, unless it was impossible.

There have been many restrictions applied by the private sector. The financial investors have often restricted services and equipment to one country or source, even to the point of reserving certain types of services and equipment, or sales of products, to either themselves or to parent companies. For example, a buyer's credit arranged with the assistance of a package supplier may give him total control over sources of equipment and supplies procured from abroad; or a joint-venture project with local partners in (say) civil engineering may contract the civil works to one of its shareholders; or a joint-venture with a technology and buy-back agreement may give to the foreign partner full control on imported inputs and sales of products abroad.

8.3.2 Joint projects with multiple objectives

Whether a project has been financed by external assistance or the private sector or not, its profitability can be expressed in several ways. For example, the profitability of a project financed by a capital grant for technical assistance involving research and development should be increased production or improved standards by producers intended to benefit from the results, even though they are not necessarily linked to the work in any way; that of a loan for a national or regional project should be internal rates of return on investments for commercial ventures generated or supported by the project; and that of private investment should be direct financial returns to the investor.

However, most projects have involved both the international and national public sectors and the private sector so that the profitability has been expressed not only in financial terms but also in ways which fulfil economic and social objectives of the administrations concerned. For example, projects concerned with research and development have been financed mostly by capital grants for technical assistance from international aid agencies. Decisions with regard to such projects have often combined specific requests presented by the host country with the particular programmes and views of the donor towards the recipient. On the whole, the profitability of a research and development project has been more its capacity to produce technical advances from which will be derived economic benefits.

Many commercial projects, on the other hand, have been financed by credit from the regional development banks at both national and regional levels. They usually have included components for services (including foreign expertise), capital investment (including foreign equipment), and a line of credit for small-scale farmers, and even companies. Within these components individual sub-projects have been identified and evaluated, each on the basis of estimated feasibility and internal rate of return (IRR). This is carried out in the same way as for the total project at the time the loan was agreed, and depended on proper implementation of previous projects by the recipient country and of the projected IRR.

However, additional aid to commercial projects has often come in the form of special services and/or equipment. These have been provided by an international aid agency, the government, or certain non-government channels, all of which can participate in commercial projects if the IRR is initially too low without some form of non-reimbursable assistance. The decision to contribute aid to such a project has invariably been based on its economic benefit to the area and/or beneficiaries, combined with the financial acceptability of the loan-financed part of the project.

Such additional aid should not be a part of any commercial project, which should be capable of remunerating the invested capital, but in reality this has not been the case. A number of countries have subsidized aquaculture investment in one form or another, through direct non-reimbursable grants on structures, by various forms of support in the acquisition of foreign technology, and by extensive exemptions of taxes and customs duties (see Chapter 7). On the whole, these are incidental to the overall costs of the project and all decisions for a commercial project are essentially based on financial returns.

There have also been variations in projects which have elements of both examples noted above. For example, commercial projects in centrally planned economies have many aspects in common with local or regional development projects, thus benefiting from considerable direct governmental aid. Projects using farms for demonstration, which may be entirely or partially owned by shareholders, and include training components, have aspects of all variations described above in financing as well as decision-making.

Chapter 9
Monitoring and Evaluation

Synopsis

This chapter is concerned with the last management tool available to the government for sector development. It summarizes the approaches and methodologies currently used for monitoring aquaculture development activities and evaluating the results. For activities at the sector level, which concern government policy and policy instruments, it notes the individual role and work of the agency responsible to policy-makers. For programme and project monitoring and evaluation it describes the responsibility of those involved and, through illustrations, lays out structured approaches currently used by the international agencies. The chapter concludes with monitoring in private investments, and summarizes the many aspects of risk management.

Monitoring (and to a lesser extent evaluation) of any activity is a valuable tool for management. For managing activities concerned with development it is subdivided by the two levels at which it is used, namely:

- the sector (organization and management) level, and
- the project (industrial) level.

Activities at the sector level are the national (or state) government policy instruments as described in Chapters 6, 7, and 8. Those at the project level are the individual investments (or grants) into production, marketing, research, training, etc., made by either private enterprises or public agencies. Reference to the importance of monitoring, reporting, and control in the planning process have also been made in an earlier chapter (see section 4.4).

Monitoring is an important management element of all plans, programmes, and projects concerned with sector development. Evaluation, on the other hand, is a management tool created and used mostly by the international financial institutions and assistance agencies in association with their development programme and project activities. However, both are formal and structured processes.

Definitions of monitoring and evaluation

Monitoring – is the continuous oversight of any activity during its implementation to ensure that operations are proceeding correctly.
Evaluation – is a learning and action-oriented management tool and process for determining as systematically as possible the relevance, effectiveness, and impact of activities in the light of their objectives, in order to improve both current activities, and future planning, programming, and decision-making.

- On-going evaluation occurs during the implementation stages of an activity,
- Ex-post evaluation is performed after an activity is completed.

Source: United Nations, 1978.

Monitoring is conducted to ensure that procurement, delivery, and installation of all inputs are proceeding as planned, standards and schedules are being maintained, and that progress is being made towards achievement of outputs and utilization of results. If not, corrective measures can be proposed and taken. Evaluation, on the other hand, is not a decision-making process in itself, but provides only information for decision-makers involved with activity. It is not concerned with the routine operations of the activity, and is neither an inspection nor review.

While the principle of monitoring and evaluation is well-recognized by all those concerned with aquaculture development and planning, methodologies to execute them differ from country to country, and agency to agency. The definition and use of various terms to describe activities and processes can also differ. This is because the international financial institutions and assistance agencies, which have been responsible for introducing these management tools and formulating their methodologies for themselves and for governments, have different goals and objectives for their programmes and projects. For example, (as noted in section 8.2) the multilateral development banks provide concessionary loans or joint-venture with equity participation, and therefore they are concerned with direct capital investment in aquaculture facilities and economic growth which should follow; the multilateral and bilateral donors, on the other hand, provide capital grants, and consequently are more concerned with development which may be institutional strengthening, technology transfer, research, social and cultural improvements, education and training, and improving the aquatic environment – all of which may eventually have some impact on economic growth but not directly.

Ideally, monitoring and evaluation indicators and techniques are established as part of the formulation of any aquaculture development programme or project. These are the specific measures of change or results as expected, and are designed to provide a scale to measure progress towards the principal objectives. For sector development they are identified with the effectiveness of the existing policy instruments; for development projects involving financial investment loans they are determined by indicators such as internal rates of return, use and repayment of loans, marketing costs, increased consumption, etc.; for projects based on capital grants, useful indicators are more difficult to identify and measure, and frequently no more than a schedule is established for monitoring and evaluation, leaving the evaluation team to set its own criteria.

9.1 Monitoring at the sector level

Monitoring of the aquaculture sector is normally the responsibility of an administrative department within the ministry given the mandate to organize and manage it; this is usually the ministry of agriculture, fisheries, or rural development. However, some selective monitoring may also be carried out by a department or institute responsible for government monitoring and evaluation of all national sectors, and located within the ministry of finance, or even the prime ministry.

Monitoring is mostly an internal task of the respective ministry concerned with aquaculture, predominantly reviewing the effectiveness of the selected policy instruments, and overseeing the administration and operation of a programme of work and assembly of projects mandated and budgeted in the current five-year development plan. However, through the five-year plan, the programme of work and projects are linked back to broad national policies for such things as food price stability or increased employment. These linkages open up opportunities for external monitoring by an inter-agency group. Unfortunately, the reality is that external monitoring meetings are very rare, as invariably they become sessions where individual mandates have to be protected, or the government's macro-economic policies always take precedence over policies of a food agency, which are usually built around production targets.

Monitoring at the sector level is almost entirely carried out by economists and analysts of the department or institute (as noted above) within a senior ministry responsible for national development, such as the ministry of finance or possibly within the prime ministry itself. The role of this institute is principally to provide the government with information which will indicate if current policies are working, and policy instruments are being effective. If not, then the information assembled by the institute should be an adequate baseline for changes to be made by the policy-makers.

The most difficult task of a monitoring institute is to analyse the effectiveness of selected policy instruments. Problems arise for a number of reasons, some of which may not be within the control of the government. For example, after 1991, the demand for smoked Pacific salmon in the USA dropped with the importation of cheap smoked Atlantic salmon from Norway and Scotland. The broad importation policies could not change the prices of the imported product, and the solution was to encourage national products to become more competitive through improved quality, increased productivity, more attractive packaging, and market promotions.

Another problem occurs if policy instruments adopted are in conflict. This often arises when a large number of instruments are in use throughout the depth of the industry. For example, in the early 1980s, several Asian countries had government credit schemes for producers of marine fish in coastal ponds with the objective of increasing production for local markets; other schemes supported production and processing of marine shrimp for export markets to earn foreign exchange. Consequently, many fish farmers converted their ponds to shrimp production. Conflicts invariably concern employment, where efforts to increase efficiency on the farm also affect local employment. This situation is often alleviated by the increased employment in processing jobs, in response to increased productivity and efficiency, but it is not always the case.

Finally, analysts have to monitor the effects of the policy instruments over time. For example, government policies for the massive increased production of marine shrimp in Indonesia and the Philippines throughout the 1980s were set back later when other countries (such as China and India) came on stream, and world prices fell.

Thus, any single instrument, or group of instruments, can never be considered the optimum to pursue a specific policy or cure a specific problem. Policy instruments have to be realistic for the time, and subject to change in the future. It is the responsibility of the government's monitoring institute to have the necessary information to advise the policy-makers that the existing instruments are no longer achieving the desired policies, and changes should be made.

Information for policy-makers, however, cannot be provided by any single monitoring or evaluation study. Consequently, the institute continuously carries out evaluations of a host of sector activities to provide a comprehensive perspective for policy thinking for aquaculture (and invariably relevant to several other sectors at the same time). For example, in a country which is planning to double its output of marine shrimp for earning export revenues and increasing domestic consumption, over a period of five years (usually the length of any national plan) evaluations might be made of:

- the effect of government exchange rate policy on the revenues and costs of shrimp farming,
- the effect of government intervention policies (such as subsidies and price controls),
- the economic benefits of a government marketing system,
- changes in the purchasing power of urban communities, and
- regulation of transport and marketing channels.

In addition, the institute may commission studies by other organizations and assemble any relevant reports and documents to add to the information base to present to policy-makers.

The mandates of these government institutes are basically to assemble the necessary information for the policy-makers to make decisions, and they themselves do not create policies. However, the organizational and management structures in some countries enables them to be very influential in making policy, thus elevating their stature in the bureaucratic hierarchy. Consequently, their monitoring and evaluation may not be as impartial as it should be, and they may even ignore certain facts; for example, price controls of fish (and other food commodities) in many poor countries (such as Nepal) have encouraged illegal cross-border trade (with India).

All these activities are far removed from the practical day-to-day operations of the aquaculture industry. Therefore, good policy decisions and realistic policy instruments depend on the availability of good statistics and information, together with economists and analysts who have some practical understanding of the workings of each specific trade and industry, and are not theoretical modellers. They also have to be experienced to be able to weigh one policy and a set of policy instruments against another, as many will conflict both between and within sectors.

9.2 Monitoring and evaluation of development projects

The monitoring and evaluation of aquaculture development projects is usually carried out both internally and externally. Internal monitoring is the responsibility of the project manager and the director of the respective government department in the ministry of agriculture or fisheries (or aquaculture division in the case of a private company) implementing the project. External monitoring is the responsibility of a project steering committee (or similar body), chaired by the same director, and made up of colleagues from the same department (or company division) and the project manager.

Internal evaluation is the responsibility of a secretariat, rather than a project steering committee, made up of the director and other senior government (or company) staff from a cross-section of relevant ministries and departments (or company divisions) – such as agriculture, fisheries, economic and rural development, finance, etc. The director may or may not choose to chair the meetings. External evaluation is the responsibility of a secretariat, but this time made up of all agencies and organizations actively involved with the project (and others who may be co-opted to serve or represent other bodies). Typical external agencies, other than national ministries, are the multilateral banks, United Nations organizations, and bilateral organizations. If the project is large and important, the meetings of the secretariat may be chaired by the minister of agriculture or fisheries (or chief executive officer). Because of the seniority of the participants, external monitoring is less frequent than internal monitoring, but nonetheless should be carried out with regularity.

The principal tools of monitoring are monthly, quarterly, and annual progress reports prepared by the project manager; monthly meetings of the project manager and the director, and regular meetings of both with the steering committee. For evaluation, the tools are mid-term or end-term evaluation reports commissioned by the secretariat, followed by meetings of the internal or external secretariat as appropriate. The evaluation reports are usually prepared by a team of outside and impartial technical experts contracted for the purpose, and representing each of the principal agencies involved.

Evaluation reports are prepared according to specific guidelines which have been laid down by respective agencies, and presented in a specific outline (see Figure 9.1).

I Executive Summary

II Project Concept and Design
- (a) Context of the project
- (b) Project document
  - the problem and technical approach
  - objectives, indicators, and major assumptions
  - beneficiaries
  - work plan

III Project Implementation
- (a) Activities
- (b) Quality of monitoring and backstopping

IV Project Results
- (a) Outputs
- (b) Immediate objectives
- (c) Development objective
- (d) Unforeseen effects
- (e) Sustainability
- (f) Follow-up

V Findings

VI Recommendations

VII Lessons Learned

VIII Annexes

Figure 9.1 Evaluation report outline used by the United Nations System.

In brief, the ex-post evaluation sets out to do three things:

- identify the results of the project, and compare them with the specific objectives as intended,
- determine the effects (i.e. the use being made) of the results, and finally,
- estimate the impact, or long-term sustainability, of the basic goals of the project activity as a whole.

Most programme or project evaluations satisfactorily complete the first two of these items but, because of the long-term nature of creating impact or sustainability, the third item cannot be estimated with any thoroughness if the ex-post evaluation is made immediately at the end of the project.

As impact, or sustainability, is the principal objective of any development initiative, evaluations are often carried out several years after project completion to determine this key indicator of project success. These may be carried out on selected projects or on a group of projects which had a common theme.

The thematic evaluation of aquaculture

Between 1984–86, UNDP, FAO, and the Norwegian Ministry of Development Cooperation financed a thematic evaluation of aquaculture specifically to determine the impact of international assistance provided to the aquaculture sector in developing countries. The evaluation was carried out on 114 aquaculture development projects funded by UNDP and executed by FAO between 1972 and 1984, of which 53 were subsequently evaluated in greater depth in the field.

The evaluation determined that:

- several projects aimed at supplementing food production in rural areas have led to improved nutritional standards and better living standards for the producers, although the numbers of producers involved and quantities produced frequently have been lower than planned.
- impact is notable in Africa, but mostly confined to Cuba among efforts in Latin America and the Caribbean. Improved and expanded practices have resulted in Indonesia, Malaysia, Nepal, Sri Lanka, and Thailand.
- activities have not led to deterioration of the environment or increased pollution, and private investment following international assistance has grown. No projects were aimed at improving the situation for women through aquaculture, nor with distribution of benefits between the sexes, but there were no indications that either sex has been favoured.

Source: FAO, 1987.

9.3 Monitoring private enterprises

Investment enterprises in the aquaculture industry are made to make a profit, and all operations and management decisions are directed towards that end. The welfare and survival of the aquatic animals or plants produced by the project to generate income from sales are therefore the most important priorities for the investors. However, the husbandry and production of aquatic animals and plants is the combination of several biological processes, none of which can be expected to achieve their predicted ends with a high degree of statistical certainty. The biological risks to any aquaculture investment are therefore too numerous and varied to list, or to describe by any order of magnitude, or to predict their frequency of occurrence.

Aquaculture production is also exposed to the impact of several natural physical and chemical processes (see section 2.1), such as earthquakes, floods, and environmental pollution. Fortunately most of these processes only occur over prolonged periods of time, and they have the appearance of stability and predictability. Nonetheless they do occasionally occur and, together with the failure of many biological processes, they contribute to the classification of aquaculture as a 'high risk' industry, similar to many other agricultural industries, such as animal and cereal production, and market gardening.

Unfortunately, the investor is not limited only to risks which occur during production. Present and future profitability of any project are also dependent on post-harvest handling and marketing of the product, and the preparation of the product by the consumer, both of which are invariably outside the investor's direct control. It is therefore important for the investor to be aware of the economic dimensions of all the risks to the farm product from birth to final consumption, and select the most appropriate management technique to avoid or minimize them.

High risk industries like aquaculture are not necessarily to be avoided. Most produce opportunities for a rapid and high return on investment, provided that the investors maintain a constant regard to all the production operations. A technique for constantly monitoring the investment, and its risks, is called 'risk management'. A detailed document on *Aquaculture and Risk Management* has been produced by Secretan and Nash (ADCP/REP/89/41), from which the following sections have been summarized.

The process of managing risk is based on the individual analyses of three fundamental activities, taken in sequence, and syntheses of the results into a programme of management action. These are:

- *Identification of risk*, or discovering the source(s) of risk from which a potential risk may arise,
- *Measuring risk*, or evaluating the impact on an individual or an organization in the event of a potential risk occurring, and
- *Managing and controlling risk*, or selecting the most effective method(s) to deal with a potential risk.

These three components have many sub-components, discussed in the following sections.

9.3.1 Identification of risk

In spite of the great diversity of aquaculture (see section 1.2), and the systems and practices used for the production of about 150 species of aquatic animals and plants world-wide (see section 1.4), the risks to any enterprise can be identified within a simple framework. All risks are fundamentally separated into:

- Business risks,
- Pure risks, and
- Management risks.

These are sub-divided as shown in Figure 9.2.

BUSINESS RISKS	
Production Risks	
Operational risks	Risks which interrupt the production cycle, such as mechanical failure, failure of technical processes, late delivery of supplies and services
Technological risks	Risks associated with a lack of adequate technology, such as hatchery propagation, or lack of technical information and expertise
Financial risks	Risks due to government financial policies, use and dependence on government policy instruments, terms of credit, changes in operational costs
Social risks	Risks due to actions of special interest groups, such as environmentalists and conservationists
Market-related Risks	Risks due to loss of product quality, lack of market information, actions of third party (the marketing middlemen)
Consumer-related Risks	Risks due to loss of consumer appeal, health regulations, actions of third party (the consumer)

PURE RISKS	
Physical risks of nature	Risks due to extreme climatic and meteorological conditions (wind, flood, drought, earthquake, volcanic action)
Social and political risks	Risk due to theft, malicious damage and fraud
Liability	Risks due to legal actions against the farm operations

Figure 9.2 A framework of the principal risks to aquaculture (Source: ADCP/REP/89/41, FAO, Rome).

Business risks

Business risks are directly related to the production of aquatic animals and plants, and the associated commercial business of the industry. These are sub-divided into the three principal product processes, as noted above, namely production on the farm, processing and marketing, and preparation for consumption.

Production risks are the principal concern of the producer. They can be divided into four categories – operational, technological, financial, and social risks. Operational risks are defined as those which disrupt the production process. These are mostly mechanical in nature, such as the failure of life-support systems (water and oxygen/air delivery, electrical power), or of structural systems (anchorages for floating cages, pond berms or bunds). Such risks are reduced by regular maintenance, together with mechanical alarm systems in support of 24-hour staff vigilance. Also included in this group of risks are poor handling of the stocks (during routine movement and sampling), and unavailability or untimely delivery of resources, such as seed, feed, and fertilizer.

Technological risks include the lack of technical information to circumvent problems arising in the production processes. Because of aquaculture's relative novelty, producers are very susceptible to these risks.

Financial risks are common to most business enterprises, and may be categorized with 'pure risks'. However, again because of the newness of the sector, financial risks to aquaculture production are acute, primarily because of lack of information and experience by the banking industry and the government in organizing appropriate credit and fiscal measures. Also included here are the consequences of sudden price changes for necessary resources, such as feed, and market prices.

Social risks include the competition for production resources, such as water and land, by other groups, together with the growing public resistance to anything which might change the ecological *status quo*.

Production risks can be alleviated by livestock insurance, which is a proven method of diverting risk. However, as the risks are so diverse, many underwriters often only agree to share the risk, and restrict the cover in some way.

Market-related risks are those to which the product is exposed once it has left the farm-gate, and which may reduce its value. These include poor processing, and the markets identified by the middlemen for final sale. These risks to product quality and value are usually alleviated by producers working closely with reliable and efficient processors and middlemen, and keeping themselves well informed about this part of the industry.

Consumer-related risks are typically outside the influence of the producer. The greatest risk to the consumer is contamination of the product which exposes consumers to ill-health. This can only be reduced by making certain that the product is grown in unpolluted waters, and processed and handled hygienically. This can be assisted by packaging the product with directions for handling and preparation, and marketing the product through reliable marketing associations.

Pure risks

Pure risks are not specific to the aquaculture industry, but to life and industry in general. Primarily pure risks include all physical risks of nature, social and political risks, and liability.

The *physical risks of nature* include external forces which damage the hatchery or farm, with the loss of infrastructure and/or the crop. Typical physical risks include floods, drought, storms, and earthquakes. Consideration for the potential of these risks occurring is taken into account during site selection, but even then there is no guarantee that nothing will happen.

Similarly *social and political risks*, such as theft, malicious damage, fraud, social unrest, and possibly war, are considered during the siting and the construction of any aquaculture infrastructure, but are again not necessarily avoidable altogether.

Liability is a pure risk which can usually be avoided with foresight and good management. Primarily it concerns the health and safety of the workers on the farm, business clients, and the public at large. Unfortunately, with the newness of the aquaculture industry, problems are not always easy to anticipate; for example, the use of many construction and cleaning materials on farms and in hatcheries has proved to be toxic to aquatic animals and plants. Liability risks, on the whole, can be reduced by high standards of all practices in the industry.

9.3.2 Measuring risk

Measuring, or quantifying, specific risks is important to place them in an order of priority for decision-making. The two elements associated with any risk are:

- the frequency of occurrence, and
- the cost and economic consequences if it occurs.

To date, such fundamental analyses have not been practised by aquaculture managers and producers in the new industry, causing many enterprises to close for reasons which could have been avoided.

The principal decisions faced by any investor, whether in production or any of the other supporting industries are:

- *Commercial decisions*, based on the financial comparison of the anticipated return on investment and the cost of any risk.
- *Mitigation and control decisions*, specific to any risk for its total elimination or reduction to an acceptable level, and at what cost.

Environmental Data	
Climatology	Basic weather data, and incidences of extremes
Hydrology	Basic physical data of waterbodies (range of tides, wave direction, pitch, etc.), water chemistry, and all seasonal changes
Geology	Topography, soil composition, and chemistry
Biological Data	
Species data	Life history cycle, basic physiology, reproduction
Species pathology	Specific diseases, incidence, treatment, efficacy of treatment, known epidemics, regulations regarding diseases
Aquatic biology	Plankton profile and seasonal blooms
Production Data	
Carrying capacity	Stock densities
Feeding	Feeding rates, feeding behaviour
Harvest	Size, methods, handling
Engineering Data	
Site works	Standards and codes of practice for facility construction (and tanks, cages, rafts), water systems, moorage
Operations	Alarm systems
Social Data	
Employees	Regulations for health and safety, working conditions
Non-employees	Local conditions, level of unemployment
Economic Data	Costs of design services and construction, operating costs, marketing data, production profiles, internal rates of return

Figure 9.3 Checklist of information required to measure risks (Source: ADCP/REP/89/41, FAO, Rome).

- *Financing decisions*, which deal with ways to finance the risk, such as insurance, and their acceptability.

In order to make any of these decisions, the investor requires considerable background information and numerical data, some of which may, even now, not be available. A check list for the typical information required in the principal areas of aquaculture enterprises is given in Figure 9.3.

9.3.3 Managing and controlling risk

There are three main techniques for the investor to manage and control risks. These are:

- to absorb the risk himself,
- improve standards and codes of practice, and
- divert the risk.

Absorbing the risk requires positive action to have a business plan and management strategy to overcome any of the business risks and pure risks (as described in section 9.3.1) as they occur. This means maintaining some financial liquidity in the business, a strategy which has not been widely followed by producers.

Improving standards and codes of practice in the industry assist in controlling many risks. However, this requires organization by the investors, and support of professional associations and groups in the industry. Typical problems which continue to face the industry are the limited number of good sites for aquaculture enterprises and the increasing use of marginal sites, the lack of pilot-scale operations before full investment is made, poor engineering standards caused mainly by investors trying to reduce capital costs for reliable infrastructure, poor advisory services provided by inexperienced private consultants and government extension services, failure to take advantage of modern vaccines and other disease controls, and the lack of concern for the health and safety of the employees.

Diverting the risk assists any investor in absorbing any loss. This is done through livestock insurance schemes, which may be private or operated by some governments in the interest of protecting an emerging and important industry.

In conclusion

Monitoring and (to a much lesser extent) evaluation are practical and valuable management tools for all those involved in aquaculture development, at whatever level. However, effectively carried out, both carry an operational cost, which may be substantial.

At the sector level, where monitoring is a function of an institute or department which is an integral part of the government structure, it becomes a line-item within the annual budget. However, as aquaculture is invariably a small and unimportant sub-sector within a larger food agency (agriculture or fisheries), then monitoring aquaculture policies and policy instruments may also receive low priority status.

In development projects, both monitoring and evaluation are usually given low priority in the planning and budget processes. Moreover, as they do not occur early in the development of any project, they may finally be neglected totally if the project over-runs its budget. In private investment projects, monitoring (or risk management) invariably has a high priority in principle but is rarely structured into the organization and management functions, and therefore is not completely applied in practice. The majority of enterprises continue to practise 'reactive management'.

The value of the management tools of monitoring and evaluation is yet to be fully appreciated by government administrators and the multitude of owners and managers in the private sector, at all levels. This attitude can best be changed first by the federations and associations of producers and tradesmen, and societies of professionals, which can encourage the practices first among their members, who are also responsible for management, before tackling administrators and policy-makers. Furthermore, more effective monitoring at the grass roots simplifies compilation of an accurate information base for the government economists and analysts, and consequently more effective policies for the aquaculture sector in response to its real needs.

PART III
PROJECT PLANNING AND MANAGEMENT

Chapter 10
Project Planning and Management Processes

Synopsis

The final chapter provides a step by step guideline to the process of project planning and subsequent management. It describes the several components of the project cycle, specifically project identification, preparation, and appraisal, in a series of 12 steps, ending with a brief review of potential problem areas.

As noted and outlined in Chapter 4 (section 4.2), the key to the formulation of a good sector plan, once development objectives are set, is a complete sector survey, an analysis of the information complete with various projections for the future, and the setting of real and attainable targets with the resources available. The targets are the specific ends to reach the required objectives. The means of achieving the ends are the overall strategy for development, specific policies defined, and selected policy instruments.

Although every one of the three policies outlined overleaf by the government of Turkey is vital to the successful sector plan, the specific projects required to achieve the ends are the most important. Moreover these include *all* projects, namely those committed by the public sector as well as those to be carried out by the private sector, which are encouraged through increased lines of credit or financed independently by investors.

Project formulation and implementation, when carried out properly and in complete detail, is a structured process, with many phases. In many simple projects the phases are short and quickly passed, perhaps in parallel with others; in large and complex projects some phases may be long and the process can only be continued when one phase is totally completed. Aquaculture project formulation has been described in complete detail, with illustrations applied, by Insull and Nash (1990). The following sections summarize the formulation and management of individual projects from this work. The information should also be placed in context with the management of development and investment projects (risk management) described in Chapter 9.

Summary of an aquaculture sector plan

Turkey has a national objective of compensating for its declining fisheries resources of the Black and Mediterranean Seas through its inland and coastal fisheries. This is to be achieved through a single strategy encompassing both its inland fisheries and aquaculture. The strategy consists of three policies, each directed towards specific aims, namely:

- enhancement of its fisheries in inland lakes and artificial reservoirs,
- expansion of its traditional small-scale entrepreneurial freshwater fish farming, and
- promotion of all forms of coastal aquaculture.

These policies are derived through policy instruments which include:

- legislative changes,
- provision of government support services, and
- projects.

Typical projects include a programme of credit for investors, with favourable terms, construction of coastal landing and operational sites for offshore farms, and construction of hatcheries for culture-based fisheries. Government activities in support include: establishing regional centres for research and development, creating a fisheries statistical service for management use, improving the extension service, and establishing a comprehensive information service, including marketing news.

10.1 Organization for project formulation and management

A project is simply a scheme to organize the use of a given quantity of resources in a specific way to achieve a particular result, all within a definite time frame. It has a precise beginning and a precise end. Usually it requires the mobilization of different skills and resources to achieve the intended result, but not necessarily so. Any small task, such as a marketing study required by an entrepreneur or the publication of a business newsletter, constitutes a project as important in its own way as the design and construction of an industrial hatchery, or a major government development project financed by a multi-million dollar loan from the World Bank. Each must be organized and managed efficiently, and completed within budget.

Consequently an idea for a project can come from any source. It may be a direct result of the sector development plan, the financial grants' policy of a technical assistance agency, or the direct need of a commercial enterprise, large or small. However, not all ideas, in spite of their individual merits, may fit directly into an overall sector development plan or a private company's plan for growth. Those which do not integrate fully with the plan have to be excluded, otherwise conflicts arise and resources are dissipated.

Project organization grows in complexity with project size and diversity. All project sponsors have therefore to estimate their capabilities to formulate and implement their projects successfully. Those sponsors with in-house capabilities, and depending on the size and diversity of the project, have the option to:

- undertake complete responsibility themselves,
- partially delegate responsibility to experts, or
- create a special task force.

Those sponsors without in-house capabilities delegate total responsibility to a specialized company, or agency.

Complete management responsibility for a large and diverse construction project is feasible when the project sponsor, such as a government ministry or a large vertically integrated private company, has its own technical study and engineering design offices, and construction and operational management units familiar with the type of facility required. The project can then be integrated into the regular organization and management operations. This model frequently exists in the agriculture sector, where public or semi-public development boards (or parastatal authorities) often have such capabilities, but is less typical of the aquaculture sector at the present time.

Project organization through *partial delegation of responsibilities* is more typical in the aquaculture sector. In this situation the project sponsor is able to formulate, implement, and operate the project, but lacks the specific technical knowledge required. It acquires this knowledge through the use of independent consultants, or consulting firms, and issues contracts for engineering and all other specialized work required. In this case the consultants are only advisers, and contractors carry out specific instructions. All key decisions remain with the project sponsor. Many capital projects in the private sector (for farms, hatcheries, and processing plants), and most technical assistance projects financed through international agencies (for national research and development centres) fall within this category.

A *project task force*, created within the organization of the project sponsor, is characteristic of large development projects within the public sector (such as regional development schemes), or investment in new fields of activity within the private sector (such as offshore farming from platformed rigs). During project formulation, or at the beginning of project implementation, a specific task force is established from existing staff who are assigned to the project for its duration, together with additional staff recruited from outside as required. The task force is usually concentrated with 'total packages' in national or international contractual tenders for activities such as project engineering, construction works, management, and technology transfer. For projects in the public sector, certain administrative rules have to be followed, ensuring that monitoring, management decisions, and operational levels are independent of each other.

Where the project sponsor does not have any capabilities in project formulation, implementation, and management, it has to *contract functions to a specialized company*. In this case the project sponsor limits its role to financing and supervision of what can become full 'turnkey' contracts; these may be extended to include not only project formulation and implementation, but also management supervision. Such arrangements are similar to those used, for example, by some hotel management companies which build and operate hotels they do not own. It is a relatively new concept in the aquaculture sector, but in the private sector such turnkey contracts are increasing and are now included in the strategies of development banks.

10.2 Stages of the project cycle

A project, from its beginning to end, is recognized to pass through five stages. These stages constitute the 'Project Cycle' (Figure 10.1). The activities within each of the five stages of the project cycle are summarized as follows, and obviously scaled to the size and diversity of the specific project proposed.

Project identification. Here the project idea is translated into a preliminary description of the project. Terms of reference for the project reconnaissance team are established, and analyses of existing situations are made. These include a broad evaluation of the future 'with' and 'without' the project. The extents and limits of the project are also proposed. Different approaches to the project are identified, and a judgement made regarding which option should be taken forward to the next stage (project preparation).

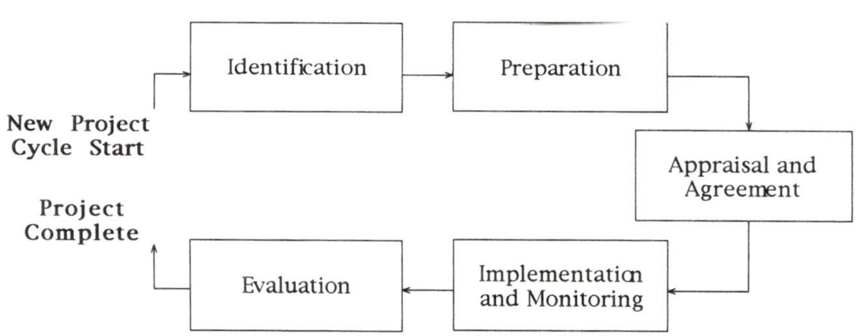

Figure 10.1 The Project Cycle (*Source*: FAO, 1990).

Project preparation. In this stage the project is designed. Objectives, prerequisites, inputs, outputs, organization, participants, clearances (administrative approvals by the government or the private client), are all defined, costs and earnings calculated, a financial plan is prepared, expected results are analysed, the socio-economic and environmental impacts are estimated, and the provisional and final project documents are prepared.

Project appraisal and agreement. Appraisal documents are prepared from the project documents and a succession of appraisal meetings, clearances, and financial negotiations take place. This brings the project to the point of meeting the required start-up agreement conditions, sometimes after revision and adaptation of project schedule, cost, objectives, and financing.

Project implementation. Here, the project management and lines of command are established, and various implementation procedures laid down. In the course of implementation project progress is monitored, revisions and adaptations are made for unexpected events, and finally the project is brought to completion.

Project evaluation. This stage takes place at a suitable time after the project has been implemented, and project objectives, project implementation, and project benefits are appraised. This evaluation may result in the project being extended, or in the identification of a new project, and may lead to the revision of methods by which similar projects will be formulated in the future.

There are other types of evaluation which may be carried out earlier in the project. These include 'project completion reports', which seek to identify problems which have arisen during implementation and which might have been foreseen, and 'mid-term' evaluations, which are characteristic of technical assistance projects.

10.3 An overview of project formulation

10.3.1 The project idea

All projects start with an idea. Any person (or group of people) who has a reasonably well-informed idea about how resources may be invested, may be said to have a project idea. A project idea might be expressed in national sector development plans, in reports of donor agencies programming missions, in the mind of an individual entrepreneur, or part of a private corporation's development strategy. If an entrepreneur or a corporation develops a project idea, financial resource are usually available to carry out further preliminary investigations. Others have to convince the financial institutions of the validity of their project idea to carry it out further. However, in many cases, practical experience in aquaculture project planning is lacking.

A project idea may have been developed from research and analysis of extensive materials, and a close knowledge of existing aquaculture activities in the proposed project area. For example, a plantation company may own land, adjacent to a river, no longer needed for cane sugar production and would like to diversify. After some research into the subject the company decides to examine the feasibility of developing the land for aquaculture production.

Alternatively a project may have little existing background, and the idea itself may not be related to any specific species or geographical location. For example, a company in seafood retailing might seek to secure future supplies of high-value fish by investing in farmed production but does not, at that stage, have any knowledge as to where and how an investment would be most profitable.

Whatever the background may be, if it is decided to carry the idea further and to formulate a project, then a process is initiated which takes the project through the first three stages of the project cycle (see section 10.2) namely, identification, preparation, and appraisal.

During the formulation process, from identification to appraisal, the project is neither approved nor established; it simply obtains increasing degrees of recognition by all parties who may be concerned or affected by it. During this process, many objections and criticisms may be received, and the project's objectives, budget, and schedule may have to be raised and perhaps even restated before the project is accepted. The project may indeed be cancelled at any one of these stages, if it does not satisfy the criteria which have to be set. Only the last stage, appraisal, leads to the final decision whether to implement the project, with full agreement as to its components, or not to proceed further.

10.3.2 The six phases of project formulation

The three stages of the project cycle are taken in sequence, but the work can be broken down in many different ways. The constituent parts are usually called 'phases', which are further broken down into 'steps'. Phases and steps can overlap to some degree. Each step contains one or more 'tasks'.

A framework for project formulation, adapted from the FAO *Guide for Training in the Formulation of Agricultural and Rural Investment Projects* (FAO, 1986), is shown in Figure 10.2, below.

Figure 10.3 further illustrates this concept and shows the possible overlaps between phases. The activities in each phase are described in the following section.

This framework differs from that which conforms to the general practice of international donor agencies, where the project formulation team is not usually involved in appraisal. This function is usually carried out by a

Stages of Formulation	Phase
Identification	Preparation for project formulation Reconnaissance and preliminary project design
Preparation	Project design Analysis of expected results Project documentation and submission
Appraisal	Negotiating the project

Figure 10.2 The framework for project formulation.

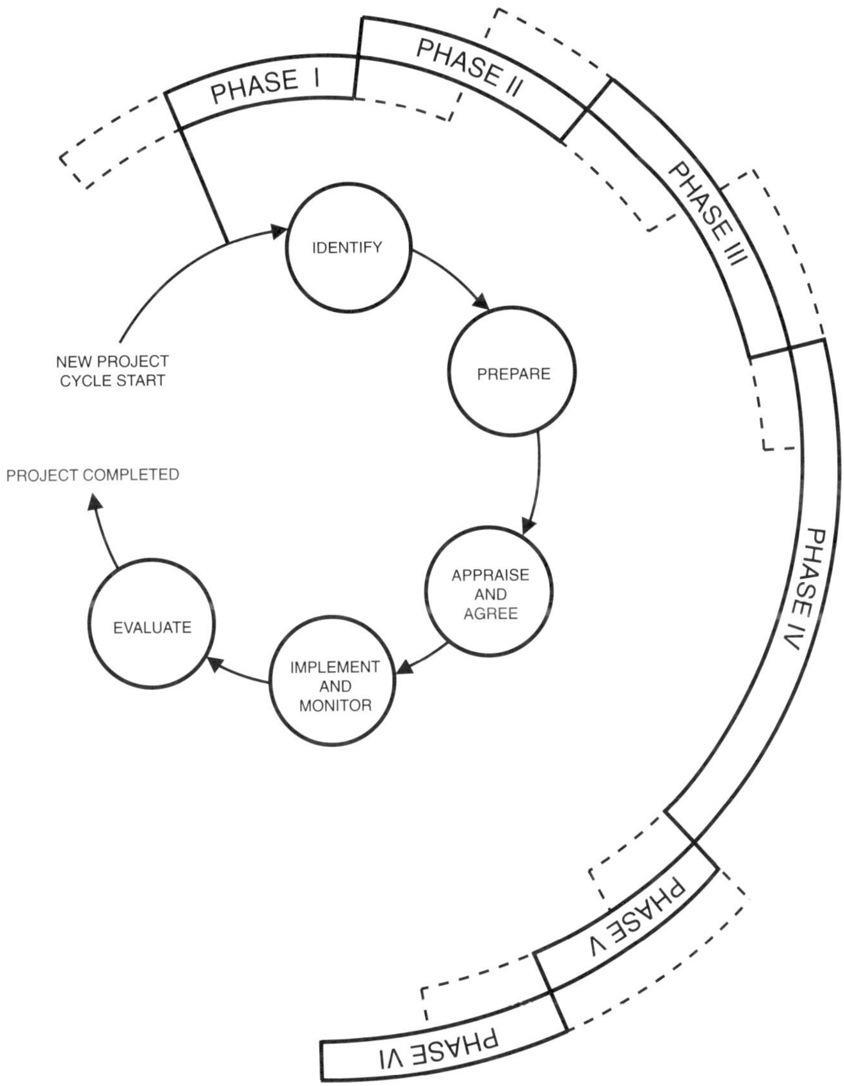

Figure 10.3 Phases of project formulation and project stages (Source: FAO, 1990).

separate group, often led by a staff member of the agency concerned. Outside the international donor system, however, it is usual for the project formulation team to continue to be involved until the implementation agreement is signed. Under this arrangement, both project appraisal and negotiation are included in the process.

The phases, steps, and outputs of the project formulation process are shown in Figure 10.4, and are described in the text of the following sections.

PHASE I: Preparation for project formulation

This phase concerns all activities necessary to prepare for sound formulation of the project. It has two steps:

Step 1 – project inception, and
Step 2 – preparation of a formulation workplan.

The output is the programme of work for project formulation.

PHASE II: Reconnaissance and preliminary project design

The second phase contains all activities necessary to define the objectives of the project, identify and consider options for meeting these objectives, making a preliminary assessment of the content of the project, and its likely effects. It has four steps:

Step 3 – analysis/diagnosis of the situation from an overall perspective,
Step 4 – analysis/diagnosis of the situation from the perspective of the main interest groups involved,
Step 5 – assessing the future without the project, and
Step 6 – outline specification of a possible project.

The output of the second phase is the preliminary design of the project, including identification of its main features, such as location, type of participants, main activities, size, timing, organizational structure, and management system. It also includes a first estimate of cost, and a tentative assessment of viability and risk. While this report may be best termed 'the identification report', it may also be called a 'project reconnaissance and preliminary design report'. In the private sector it is often called a 'project feasibility study'.

At this point the report is normally submitted to the funding organization for approval before further work is undertaken.

PHASE III: Project design

This phase normally initiates the formal project preparation stage. It typically has five steps:

Step 7 – detailed technical and socio-economic investigations,
Step 8 – more precise definition of project objectives, targets, and design criteria,

PROJECT IDENTIFICATION		PROJECT PREPARATION			PROJECT APPRAISAL
PHASE I	PHASEII	PHASE III	PHASE IV	PHASE V	PHASE VI
PREPARATION FOR PROJECT FORMULATION	RECONNAISSANCE AND PRELIMINARY PROJECT DESIGN	PROJECT DESIGN	ANALYSIS OF EXPECTED RESULTS	PROJECT SUBMISSION	PROJECT NEGOTIATION
Step 1 Project inception *Step 2* Preparation of formulation workplan	*Step 3* Analysis of the situation from an overall perspective *Step 4* Analysis of the situation having regard to the main interest groups involved *Step 5* Assessing the future 'without the project' *Step 6* Outline specification of a possible project	*Step 7* Detailed technical and socio-economic investigations *Step 8* Definition of project objectives, targets, and design criteria *Step 9* Design of individual project components *Step 10* Design of project organization structure, and management arrangements *Step 11* Project cost, revenues and financing proposal	*Step 12* Financial analysis *Step 13* Economic analysis *Step 14* Social analysis *Step 15* Environmental impact analysis	*Step 16* Project documentation and submission	*Step 17* Project appraisal and negotiation
↓	↓	↓	↓	↓	↓
Project Formulation Work Programmed	Development potentials and constraints analysed, possible project identified	Project designed in detail	Financial, economic, social and environmental impact of project assessed	Project document completed and presented	Project accepted and financed

Figure 10.4 Phases, steps, and outputs of project formulation (Source: FAO, 1990).

Step 9 – design of individual project components,
Step 10 – design of project organization, structure, and management arrangements, and
Step 11 – project cost and revenue estimation, and first financing proposal.

The output of this phase is a full description and costing of the project, together with a proposed financing plan. In external assistance projects this phase is not usually the subject of a separate report. In the private sector it may be, when it is usually called a 'feasibility study'.

A number of adjustments and possible revisions in project design will often be requested. These lead to additional work, often carried out in parallel with the next phase.

PHASE IV: Analysis of expected results

The fourth phase concerns all activities necessary to assess project results in terms of outputs, effects, and impact on the sector, and on any other sectors they may affect in some way. Its starting point is usually the project design report, which provides the basis for analysis. The work usually contains four steps:

Step 12 – financial analysis,
Step 13 – economic analysis,
Step 14 – social analysis, and
Step 15 – environmental impact analysis.

The output of the phase is the determination of effects and impacts of the project.

PHASE V: Project documentation and submission

The fifth phase concerns all activities necessary to prepare a final project document, complete with design and relevant analysis. The work of the phase may be described in one step, namely:

Step 16 – project documentation and submission.

The output of the phase is the project document.

In external assistance projects the work of the formulation team is usually complete at the end of this phase. In the private sector, however, participation of the team members may be of value in the project negotiation phase, particularly when several financing agencies are concerned, and/or if negotiations necessitate adjustments to the project.

PHASE VI: Negotiating the project

The sixth phase concerns all activities necessary to have the project document accepted, and the project financed for implementation. It starts when the source of financing accepts the project formulation document. The work includes only one step:

Step 17 – project appraisal and negotiation.

The output is a project fully ready for implementation, under proper administration, and with the necessary financial commitments.

10.4 An overview of project implementation

This section describes two aspects of project implementation, namely the sequence of the main implementation phases, and problem areas at different stages of implementation.

Implementation is described by further succession of phases, each phase containing a number of steps. However, because projects differ considerably, the process described below is of a general nature only. The phases are numbered consecutively in the following text to avoid any confusion with those in project formulation.

10.4.1 Project phasing

There are typically six phases in project implementation. These are:

Phase VII: Recruiting the human resource,
Phase VIII: Studies and engineering,
Phase IX: Construction and procurement,
Phase X: Start-up of field operations,
Phase XI: Standardization of field operations, and achievement of project goals,
Phase XII: Termination of the project component.

Projects need not necessarily include all these phases, or all in their entirety. For example, in a project to expand existing mussel production and increase regional marketing there may be four main components, namely (a) a market study for increased production, (b) an extension service for new farmers, (c) a credit programme, and (d) project management, only five of the phases would be necessary. In this situation, Phase IX (construction and procurement) would not be required, while Phase VIII (studies and engineering) would consist of the market study only, without any engineering. Also the phases would vary in the length of time they occupy in project implementation. For example, project management will begin when the project starts, and will finish when it ends, but the market study would be fully implemented early in the project schedule.

The characteristics of each phase are as follows:

PHASE VII: Recruiting the human resource

This phase begins as soon as the project agreement is signed. Key steps in this phase are:

- recruiting the project manager and project assistants, and recruiting managers and assistants for any components,

- establishing a project management office and (if needed) component offices, and providing means of operation, and
- recruiting project consultants and (if needed) consultants for specific components.

The phase ends when the last recruitment has been made.

PHASE VIII: Studies and engineering

The next phase, concerned with project studies and engineering, begins as soon as the project management office is operational. Depending on particular component schedules and characteristics, some studies and engineering work may be delayed (until start-up of the component concerned), or reduced to studies only, if no engineering is needed. Key steps in the phase are:

- site studies and topographic surveys,
- monitoring environmental parameters,
- engineering of facilities and preparation of tender documents, and
- selection of equipment and supplies, and preparation of procurement lists.

The phase ends when the last construction or supply tender/procurement document is fully completed and approved.

PHASE IX: Construction and procurement

The phase concerning construction and procurement begins as soon as the first tender or procurement dossier has been approved. Key steps in this phase are:

- selection of tenderer/supplier, and issuing of tender/supply requests,
- analysis of offers, and awards of contracts,
- supervision of contractors, works meetings, payment of contractors,
- works and supplies control and agreement, customs clearances, and
- post-delivery control and guarantees.

This phase ends when the last construction or delivery is completed or received, accepted, and terms of guarantees have been met. In construction work this usually means a full year after construction is completed.

PHASE X: Start-up of field operations

The next phase deals with field operations. It begins as soon as the first field team has been hired, and provided with necessary working tools, such as vehicles, equipment, office facilities, etc. Key steps in this phase are:

- field testing of working tools,
- meeting expected minimum biotechnical and performance standards,
- meeting expected minimum financial performance standards, and
- meeting expected minimum training, extension, and other standards.

This phase ends when the field operation has reached all the expected standards.

PHASE XI: Standardization of field operations and achievement of project goals

This phase begins when project start-up has been completed, and management can turn its attention to the main project priorities, such as selling products from the completed hatchery or farm, and/or improving the performance of project components. This phase ends with project termination.

PHASE XII: Termination of the project component

This is the last phase. It overlaps the preceding one, and begins when all the goals have been reached, and/or funds exhausted. Key steps in this phase are:

- preparation for project termination,
- project follow-up,
- staff reallocation.

The phase ends when the last administrative structure of the project has been closed down, and all the staff dispersed.

10.4.2 Problem areas in project implementation

All projects have problems, which may be large or relatively minor. The ways by which problems occur vary from project to project, but principally they fall into certain well-defined areas. These are briefly summarized as follows.

Institutional problems. These arise when authority, responsibility, or working relationships are not fully clarified, with formal and properly applied lines of communication and decision-making authority among the project teams and the organizational units. For example, in a large regional development project, difficulties in obtaining land and water for production ponds for every interested farmer may be the result of dealing with the wrong government organization, poor project management and staffing, equipment procurement difficulties, and poor monitoring and evaluation.

Conceptual problems. These arise from inadequacies in project formulation, resulting from insufficient background studies or errors in analysis, or inadequacies in the planning work to be performed. For example, in a project to introduce subsistence aquaculture to a developing country, the concepts may be too flexible, the objectives may be wrong or non-sustainable, project components may be too numerous or too large, or the schedule may be unrealistic.

Technical problems. Such problems usually arise from unexpected factors discovered during project implementation, or errors in project implementation work. For example, in a project to convert traditional fish production ponds into shrimp ponds, there might be a consistent shortfall in production of shrimp due to seasonal water quality changes and a low-quality feed not foreseen in project formulation, or the standard of engineering design and/or construction to deepen the ponds and exchange more water may be poor.

Financial problems. These arise when procedures and schedules for funds, manpower, supplies, and equipment, etc., necessary to carry out project activities have not been adequately organized, or delays in implementation result in additional costs, or project costs have been under-estimated. For example, in a large project to clear mangrove for shrimp production ponds, there may be recurrent budget shortages due to poor scheduling of heavy earth-moving equipment for digging new canals or dredging, or, as a result, under-estimated investment or operational costs. Financial problems may also occur because of sudden changes in the price of feed, or market value of the products, all as a result of competition or other market factors.

Social problems. Such problems often result from inadequacies in the analysis of social aspects of the project in formulation work, or from changes in social balances or social organization during project implementation. For example, in a project to introduce aquaculture to traditional coastal fishermen, problems may be manifested by slow adoption of farming techniques by the fishermen who may find aquaculture less attractive and more difficult than fishing, or there may be an inequitable distribution of benefits as some sites (farms) are naturally more productive than others.

Political problems. These may result from unforeseen changes in national policy or government, or sudden unexpected political events. Insufficient government commitment may result. For example, in a project to encourage coastal communities to take-up aquaculture and discourage urban drift, there may be a change in government to one opposed to financial incentives for farmers, and all subsidies and grants are cancelled. A project may also be disrupted by sudden internal strife.

Environmental problems. Environmental problems frequently arise and affect the project, or the project itself may cause environmental damage which is unacceptable. In the former case these may arise either from project-related or external natural factors unforeseen during project formulation, or from other projects which did not exist or were not planned at the time of project formulation. For example, in a project to enhance the fisheries of inland lakes and reservoirs through aquaculture, there may be sudden water pollution from new industrial projects nearby, or competition for the same water resources by urban development; there may also be general degradation of resources, or a natural disaster, such as an earthquake or a typhoon. In the second case, and project for the production of shrimps in coastal ponds may produce a heavy discharge of nutrients from the complex causing algal blooms dangerous to traditional oyster fisheries nearby, or scouring may destroy more mangroves than initially foreseen and accepted.

Internal problems. Other problems may occur which may be related to the management and operation of the project itself. These may include the human factor where, for example, the personality of one member of the management team causes minor difficulties to be exacerbated, or there may be a '*force majeure*'.

In summary, while it is the concern of the project formulation team to design the project to minimize all potential problems or risks, in any of the above areas, so equally project management must anticipate them as far as possible and minimize their impact when they occur. It is apparent from this list of potential problem areas in project implementation that many problems which arise can be foreseen at the time of project design. Satisfactory project implementation, therefore, is dependent on sound project formulation.

10.4.3 Keeping the project on schedule

Maintaining the project on schedule is the responsibility of the project manager or, in a large project, of the project management office. Effective monitoring of the project (planned during project design) assists project management in this task. In a small project the project manager will carry out his own monitoring, but a large project may justify formation of a separate group, such as a project monitoring unit (see Chapter 9).

Most projects incur cost over-runs

In an analysis of problems associated with some 70 agricultural investment/development projects prepared by the FAO Investment Centre throughout the 1970s, it was reported that 80% had over-runs of more than six months, and 40% had over-runs of more than 20 months. *Source*: The Design of Agriculture Investment Projects; lessons from experience. Investment Centre, FAO, Rome, 1989.

Maintenance of the project implementation schedule is particularly important as, more frequently than not, the direct consequences if it is not kept are cost over-runs. It is not an easy task for management to keep within predetermined time limits.

There are a number of reasons for the many over-runs in project implementation. Among the most frequent is the considerable pressure on project formulation teams, especially in commercial projects, to complete the project in advance of the schedule. Often the teams do not have the capacity to resist 'directives' from a corporate president or a government minister. For example, in a well-intentioned attempt to complete an aquaculture project in time to accommodate a breeding cycle of the finfish or shellfish being cultured (or to lose a whole year), there is a temptation for the project formulation team at the project preparation stage to accelerate the engineering design study, or shorten the construction phase, in response to a 'promise' from the minister or governor that priority for the project will be given by the department of public works for all necessary equipment and materials.

In practice, it is less simple or advisable for the implementation team to take short cuts. Relative to the cost, design and construction proceed at a reasonably uniform and standard rate (unless the project is highly complex). Without additional human and financial resources to meet an accelerated schedule, any attempt to do so invariably backfires. This causes a delay and a cost over-run, and invariably disharmony among project partners.

Delays can be frequently traced back to a combination of conceptual, management, and technical problems. Such difficulties again often have their basis in weakness in project design, in particular failure of the formulation team to ensure that tasks to be implemented are comparable to the skills and experience of project management, and time is available to carry them out.

Although schedules for project implementation established during project formulation are invariably adequate, delays and over-runs also occur for other reasons. One frequent problem is the inability of the borrowers (usually the farmers) to comply with conditions attached to loans. Hence the rates of investment into the project by the private sector are below projected participation.

Managers can frequently avoid or counter many of the problems identified above if they are familiar with appropriate management tools, such as task analysis, or, for larger projects, network analysis.

References and Source Materials

Publications of the Aquaculture Development and Coordination Programme (ADCP), 1986–89, FAO, Rome

ADCP/REP/86/23 Observations on technical assistance to aquaculture, by C. E. Nash.

ADCP/REP/86/24 Planning an aquaculture facility – guidelines for bioprogramming and design, by Carol M. Brown and C. E. Nash.

ADCP/REP/87/25 Future economic outlook for aquaculture and related assistance needs, by C. E. Nash.

ADCP/REP/87/27 Aquaculture sector development – a guideline for the preparation of a national plan, by P. D. Maine and C. E. Nash.

ADCP/REP/87/28 Women in aquaculture, by C. E. Nash, Carole R. Engle, and Donatella Crosetti.

ADCP/REP/88/30 A regional survey of the aquaculture sector in eleven Middle East countries, by P. G. White.

ADCP/REP/88/31 A regional survey of the aquaculture sector in East Asia, by Elvira A. Baluyut.

ADCP/REP/88/32 A regional survey of the aquaculture sector in the Pacific, by S. Nelson.

ADCP/REP/88/33 Planning for aquaculture development, by C. E. Nash.

ADCP/REP/89/34 A regional survey of the aquaculture sector in the Mediterranean Region, by M. Girin.

ADCP/REP/89/35 A regional survey of the aquaculture sector in West Asia, by S. C. Pathak.

ADCP/REP/89/36 A regional survey of the aquaculture sector in Africa South of the Sahara, by B. Satia.

ADCP/REP/89/37 A regional survey of the aquaculture sector in North America, by C. B. Kensler.

ADCP/REP/89/38 A regional survey of the aquaculture sector in Easter and Northwestern Europe, by H. Ackefors.

ADCP/REP/89/39 A regional survey of the aquaculture sector in Latin America, by P. Noriega-Curtis and J. Vera Rivas.

ADCP/REP/89/40 A regional survey of the aquaculture sector in the Caribbean, by P. Ferlin and P. Noriega-Curtis.

ADCP/REP/89/41 Aquaculture and risk management, by P. A. D.

Secretan and C. E. Nash.
ADCP/REP/89/42 A preliminary review of selected legislation governing aquaculture, by Anne R. Van Houtte, Nicola Bonucci, and W. R. Edeson.
ADCP/REP/89/43 Aquaculture systems and practices – a review, by Elvira A. Baluyut.
ADCP/REP/89/44 A review of extension methodologies in aquaculture, by Carole R. Engle and Nathan M. Stone.

Other references and resources

Ackefors, H., 1983. *Development of Aquaculture in Sweden*. The Swedish Council for Planning and Coordination of Research, Temdahls Tryckeri AB, Ostervala, 39pp.

Alaska Department of Fish and Game (ADF&G), 1982. *Cook Inlet Regional Salmon Enhancement Plan, 1981–2000*. Alaska Department of Fish and Game, State of Alaska, Anchorage.

Asian Development Bank, 1983. *Indonesia Fisheries Sector Study*. Asian Development Bank, Manila, 168pp.

Asian Development Bank, 1985. *Thailand Fisheries Sector Study*. Asian Development Bank, Manila, 232pp.

Asian Development Bank, 1986. *Environmental Planning and Management*. Asian Development Bank, Manila, 282pp.

Bailey, C., 1992. Coastal aquaculture development in Indonesia. In R. Pollnac, C. Bailey, and A. Poernomo (eds). *Contributions to Fishery Development Policy in Indonesia*. Central Research Institute for Fisheries, Agency for Agricultural Research and Development, Ministry of Agriculture, Jakarta, pp. 57–72.

Castillo, Silvana, Popma, T. J., Phelps, R. P., Hatch, L. U., and Hanson, T. R., 1992. *Family-scale Fish Farming in Guatemala*. Research and Development Series No. 37, International Center for Aquaculture and Aquatic Environments, Alabama Agricultural Experiment Station, Auburn University, Alabama.

Chua, T. E., 1986. Aquaculture training in Asia: Estimates of manpower requirements. *Proceedings of the First Asian Fisheries Forum*, Asian Fisheries Society, Manila.

Corbin, J. S., 1976. *Aquaculture in Hawaii, 1976*. Department of Planning and Economic Development, State of Hawaii, Honolulu, 107pp.

Davy, F. B., 1991. *Mariculture Research and Development in Japan*. International Development and Research Centre, IDRC-MR298e, Ottawa, 97pp.

Department of Planning and Economic Development, State of Hawaii, 1977. *Permits and Environmental Requirements for Aquaculture*. Department of Planning and Economic Development, State of Hawaii, Honolulu, 69pp.

Department of Planning and Economic Development, State of Hawaii, 1978. *Aquaculture Development in Hawaii.* Department of Planning and Economic Development, State of Hawaii, Honolulu, 222pp.

Dunn, I. G., 1989. *Development of Inland Fisheries under Constraints from other Users of Land and Water Resources: Guidelines for Planners.* Fisheries Circular, No. 829, FAO, Rome, 54pp.

FAO, 1981. *Agriculture: Toward 2000.* C87/27, FAO, Rome.

FAO, 1982a. *Report of the IPFC Workshop on Inland Fisheries for Planners.* Indo-Pacific Fisheries Commission, Fisheries Report No. 288, FAO, Rome, 191pp.

FAO, 1982b. *Aquaculture Extension Services Review.* Fisheries Circular No. 747. FAO, Rome.

FAO, 1986. *Guide for Training in the Formulation of Agricultural and Rural Investment Projects.* Development Policy Studies and Training Service, Policy Analysis Division, FAO, Rome.

FAO, 1987. *Thematic Evaluation of Aquaculture.* Report by UNDP, FAO, and the Norwegian Ministry of Development Cooperation. FAO, Rome, 85pp (plus annexes).

FAO, 1988a. *Aquaculture development constraints in SADCC countries of Africa.* Internal Working Paper, FAO, Rome.

FAO, 1988b. *National reviews for aquaculture development in Africa. No. 18: Ghana.* Fisheries Circular No. 770.18. FAO, Rome.

FAO, 1989a. Rural development in Europe. *Report of the Special Session of the FAO/EEC Working Party on Agrarian Structure and Farm Rationalization,* 12–15 September, Bulgaria. FAO, Rome, 177pp.

FAO, 1989b. *Reunión técnica de planificación en acuicultura.* Caracas, Venezuala. AQUILA GCP/RLA/075/ITA, Programa Cooperativo Gubernamental FAO-Italy. FAO, Rome, 94pp.

FAO, 1990. *Rural area development planning.* ESP/TMAP/21, Policy and Analysis Division, Agriculture Department, FAO, Rome, 289pp.

FAO, 1990. *Una estrategia para el desarrollo de la acuicultura: El caso de América Latina.* COPESCAL Documento Ocasional No. 6. FAO, Rome, 46p.

FAO, 1994. *Aquaculture production 1986–1992.* Fisheries Circular No. 815, Rev.6, FAO, Rome, 216pp.

Group of Experts on the Scientific Aspects of Marine Pollution (GESAMP), 1991. Reducing environmental impacts of coastal aquaculture. *Reports and Studies,* No. 47. FAO, Rome.

Hanneson R. and J. Kurien, 1988. *Studies on the role of fishermen's organizations in fisheries management.* Fisheries Technical Paper No. 300, FAO, Rome, 48pp.

Hannig, W., 1988. *Towards a Blue Revolution; Socio-economic Aspects of Brackishwater Pond Cultivation in Java.* Yogyakarta, Gadjah Mada University Press.

Insull, A. D. and C. E. Nash, 1990. *Aquaculture project formulation.* Fisheries Technical Paper No. 316, FAO, Rome, 129p.

International Association of Fish Meal Manufacturers, 1989. The role of fish meal in the diets for salmon. *Report to the International Association of Fish Meal Manufacturers,* by I. H. Pike, G. Andorsdottir and H. Mundheim. Technical Bulletin No. 24, Potters Bar, 36p.

MacPherson, N. J., 1987. *Development of marine and inland aquaculture in Greece.* Field document, FI:DP/GRE/85/002, FAO, Rome.

Ministerio del Medio Ambiente, Colombia, 1974. *National code of renewable natural resources and environmental protection,* Ministerio del Medio Ambiente, Bogota.

Ministry of Agriculture. Fisheries and Food, 1976. Endangered species – Import and Export Act, Ministry of Agriculture, Fisheries and Food, London.

Ministry of Transport, New Zealand, 1978. A guide to approval procedures, structures and works: Foreshore and navigable waters. Information Service, Ministry of Transport, Wellington, 30pp.

Ministry of Agriculture and Fisheries, New Zealand, 1983. Freshwater fish farming regulations, Ministry of Agriculture and Fisheries, Wellington.

Ministry of Agriculture and Fisheries, The Netherlands, 1988. Fishery development. Report by the Working Group on Fishery Development, Ministry of Agriculture and Fisheries, The Hague, 17pp.

Ministry of Supply and Services, Canada, 1988. Commercial aquaculture in Canada. Communications Directorate, Department of Fisheries and Oceans, Ottawa, 30pp.

Ministry of Agriculture, (undated). China. Fishery. Bureau of Aquatic Products, Ministry of Agriculture, Animal Husbandry, and Fisheries, Beijing.

Ministry for Agriculture and Food, Hungary, (undated). Fisheries in Hungary, by L. Dobrai and G. Pékh (eds). Ministry for Agriculture and Food, 91pp.

Ministry of Environment and Renewable Natural Resources, Venezuela, (undated). Environmental impact statements, Ministry of Environment and Renewable Natural Resources, Caracas.

Nash, C. E., 1979. Structure of U.S. aquaculture. *Food Policy* 4(3) 204–215.

Nash, C. E., 1991 (ed.). *The Production of Aquatic Animals I – Crustaceans, Molluscs, Amphibians, and Reptiles.* (World Animal Sciences C:4), Elsevier Science Publishers, Amsterdam, 244pp.

Nash, C. E., 1992. *Employment and manpower in aquaculture.* (Unnumbered) Human Resources, Institutions, and Agrarian Reform Division, Agriculture Department, FAO, Rome, 91pp.

Nash, C. E. and Novotny, A. J., 1995 (eds). *The Production of Aquatic Animals II – Fishes.* World Animal Sciences, Elsevier Science Publishers, Amsterdam, 400pp.

National Fishing Convention, The Philippines, 1988. *Institutional strengthening of fisheries for development and progress.* Report of the Convention, Manila, 52pp.

National Research Council, USA, 1979. *The role of the US Department of*

Agriculture in aquaculture. Panel on Aquaculture, Board of Agriculture and Renewable Resources, National Academy of Sciences, Washington D.C., 49pp.

Norsk Fiskeoppdrett, 1987. *Norwegian Fish Farming.* Norsk Fiskeoppdrett A/S, Bergen, 48pp.

Novotny A. J. and Nash, C. E. 1995. Production of salmonids. In Nash, C. E. and A. J. Novotny, (eds). *The Production of Aquatic Animals II – Fishes.* World Animal Sciences, Elsevier Science Publishers, Amsterdam, 175–238.

Office of Economic and Cooperative Development (OECD), 1989a. *Aquaculture: Developing a New Industry.* Office of Economic and Cooperative Development, Paris, 126pp.

Office of Economic and Cooperative Development (OECD), 1989b. *Aquaculture: A review of recent experience.* Office of Economic and Cooperative Development, Paris, 331pp.

Pathak, S. C. and Mohanakrishnan, P. 1987. *Proceedings of the First Asian Fisheries Forum,* Asian Fisheries Society, Manila.

Rowat, R., 1989. *Trained manpower for agricultural and rural development.* FAO Economic and Social Development Paper, No. 10, FAO, Rome.

Science Council of Canada, 1984. *Aquaculture: A development plan for Canada,* Ottawa, 34pp.

Science Council of Canada, 1985. *Aquaculture: An opportunity for Canadians,* Ottawa, 28pp.

Swedish Council for Planning and Coordination of Research, 1983. *The Environmental Impact of Aquaculture.* Temdahls Tryckeri AB, Ostervala, 74pp.

Tacon, A. G. J., Maciocci, G., and Vinatea, J. E., 1987. *National aquaculture feed surveys for aquaculture planning and development in Latin America and the Caribbean: 1. Guidelines.* GCP/RLA/075/ITA, Field document 1/E. FAO, Rome.

Tillapaugh, D. L. and J. C. Edwards, 1980. *A permit and license guide for the prospective aquaculturist.* Ministry of Environment (Marine Resource Branch), Province of British Columbia, Canada, 23pp.

Trimble, G. M., 1972. *Legal and administrative aspects of an aquaculture policy for Hawaii.* Department of Planning and Economic Development, State of Hawaii, Honolulu, 61pp.

United Nations, 1978. *Glossary of evaluation terms.* JIU/REP/78/5, United Nations, Geneva.

US Department of Commerce, 1977. *NOAA aquaculture plan.* (ed. J. B. Glude), National Oceanic and Atmospheric Administration, US Department of Commerce, Washington D.C., 41pp.

US Department of the Interior, 1983. National aquaculture development plan. Federal Coordinating Council on Science, Engineering, and Technology; Joint Subcommittee on Aquaculture, US Department of the Interior, Washington D.C., 41pp.

World Bank, 1988. *World Development Report.* World Bank, Washington D.C.

Abbreviations

ADP	Asian Development Bank
ADBN	Agricultural Development Bank of Nepal
ADCP	Aquaculture Development and Coordination Programme
EU	European Union
EEZ	Extended economic zone
EIS	Environmental impact statement
FAO	Food and Agriculture Organization
FES	Fédération Europèene de la Samoniculture
GESAMP	Group of Experts on the Scientific Aspects of Marine Polluti
GNP	Gross national product
HIDB	Highlands and Islands Development Board (UK)
HYV	High yielding varieties
IFAD	International fund for Agricultural Development
IRR	Internal rate of return
ISIC	International Standard Industrial Classification
OECD	Office of Economic and Cooperative Development
t	Metric ton
TCP	Technical Cooperation Programme (FAO)
UN	United Nations
UNDP	United Nations Development Programme
UNESCO	United Nations Educational, Scientific, and Cultural organization
US AID	US Agency for International Development
VAT	Value added tax
WFP	World Food Programme

Index